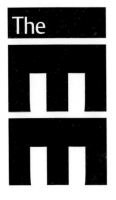
The IEE

fore the date below

se return on or b   re

Guidance Note 3
Inspection & Testing

17th IEE Wiring Regulations Seventeenth Edition
BS 7671:2008 Requirements for Electrical Installations

Published by The Institution of Engineering and Technology, London, United Kingdom

The Institution of Engineering and Technology is registered as a Charity in England & Wales (no. 211014) and Scotland (no. SC038698).

 The Institution of Engineering and Technology is the new institution formed by the joining together of the IEE (The Institution of Electrical Engineers) and the IIE (The Institution of Incorporated Engineers). The new Institution is the inheritor of the IEE brand and all its products and services, such as this one, which we hope you will find useful. The IEE is a registered trademark of the Institution of Engineering and Technology.

First published 1992 (0 85296 537 0)
Reprinted (with amendments to Section 17) 1993
Second edition (incorporating Amendment No. 1 to BS 7671:1992) 1995 (0 85296 867 1)
Third edition (incorporating Amendment No. 2 to BS 7671:1992) 1997 (0 85296 956 2)
Fourth edition (incorporating Amendment No. 1 to BS 7671:2001) 2002 (0 85296 991 0)
Reprinted (with new cover) 2003
Reprinted (incorporating Amendment No. 2 to BS 7671:2001) 2004
Reprinted (with amendments to imprint page) 2006
Fifth edition (incorporating BS 7671:2008) 2008 (978-0-86341-857-0)
Reprinted 2009, 2010

Copies of this publication may be obtained from:
The Institution of Engineering and Technology
PO Box 96, Stevenage, SG1 2SD, UK
Tel: +44 (0)1438 767328
Email: sales@theiet.org
www.theiet.org/publishing/books/wir-reg/

ISBN 978-0-86341-857-0

Typeset in the UK by The Institution of Engineering and Technology
First printed in the UK by Printwright Ltd, Ipswich
Reprinted in the UK by Cambrian Printers Ltd, Aberystwyth

Contents

Chapter 3   Periodic inspection and testing   59

Chapter 4   Test instruments   81

Chapter 5   Forms   85

Cooperating organisations

The Institution of Engineering and Technology acknowledges the contribution made by the following organisations in the preparation of this Guidance Note.

British Telecommunications plc
I. Cahill MA MIEE

Electrical Contractors' Association
H.R. Lovegrove IEng FIET
Eur Ing L. Markwell BSc MSc CEng MIET MCIBSE MCIBSE LCGI

Electrical Contractors' Association of Scotland (SELECT)
D. Millar IEng MIIE MILE

GAMBICA Association Ltd
M. Hadley
A. Sehdev BSc(Hons)

Health and Safety Executive
K. Morton BSc CEng MIEE
N. Gove MEng CEng MIEE

Institution of Engineering and Technology
J.F. Elliott BSc(Hons) IEng MIEE (Editor)
M. Coles BEng(Hons) MIEE
P.R.L. Cook CEng FIEE
G.D. Cronshaw IEng FIET
I.M. Reeve BTech CEng MIEE

Electrical Safety Council
M.C. Clark BSc(Eng) MSc CEng FIEE FIMechE FCIBSE

United Kingdom Accreditation Service (UKAS)
Eur Ing J. Pettifer BSc CEng MIET FCQI

Acknowledgements

References to British Standards, CENELEC Harmonization Documents and International Electrotechnical Commission standards are made with the kind permission of BSI. Complete copies can be obtained by post from:

BSI Customer Services
389 Chiswick High Road
London W4 4AL
Tel: +44 (0)20 8996 9001
Fax: +44 (0)20 8996 7001
Email: orders@bsi-global.com

The BSI also maintains stocks of international and foreign standards, with many English translations. Up-to-date information on BSI standards can be obtained from the BSI website: www.bsi-global.com

Illustrations of test instruments were provided by Rod Farquhar Design: www.rodfarquhar.co.uk

Thermographic surveying images are reproduced with the kind permission of FLIR Systems Ltd: www.flirthermography.co.uk

Cover design and illustration were created by The Page Design: www.thepagedesign.co.uk

It is strongly recommended that anyone involved in work on or near electrical installations possesses a copy of the *Memorandum of Guidance on the Electricity at Work Regulations 1989* (HSR25) published by the Health and Safety Executive.

Copies of Health and Safety Executive documents and approved codes of practice (ACOP) can be obtained from:

HSE Books
PO Box 1999
Sudbury, Suffolk CO10 2WA
Tel: +44 (0)1787 881165
Email: hsebooks@prolog.uk.com
Web: www.hsebooks.com

The HSE website is www.hse.gov.uk

Preface

This Guidance Note is part of a series issued by the Institution of Engineering and Technology to explain and enlarge upon the requirements in BS 7671:2008, the 17th Edition of the IEE Wiring Regulations.

Note that this Guidance Note does not ensure compliance with BS 7671. It is intended to explain some of the requirements of BS 7671 but readers should always consult BS 7671 to satisfy themselves of compliance.

The scope generally follows that of BS 7671; the relevant Regulations and Appendices are noted in the margin. Some Guidance Notes also contain material not included in BS 7671:2008 but which was included in earlier editions of the Wiring Regulations. All of the Guidance Notes contain references to other relevant sources of information.

Electrical installations in the United Kingdom that comply with BS 7671 are likely to satisfy Statutory Regulations such as the Electricity at Work Regulations 1989, however this cannot be guaranteed. It is stressed that it is essential to establish which Statutory and other Regulations apply and to install accordingly. For example, an installation in premises subject to licensing may have requirements different from, or additional to, BS 7671 and these will take precedence.

Introduction

This Guidance Note is principally concerned with Part 6 of BS 7671 — Inspection and testing.

Neither BS 7671 nor the Guidance Notes are design guides. It is essential to prepare a full specification prior to commencement or alteration of an electrical installation. The specification should set out the detailed design and provide sufficient information to enable competent persons to carry out the installation and to commission it. The specification must include a description of how the system is to operate and all the design and operational parameters. It must provide for all the commissioning procedures that will be required and for the provision of adequate information to the user. This will be by means of an operational manual or schedule.   514.9

It must be noted that it is a matter of contract as to which person or organisation is responsible for the production of the parts of the design, specification and any operational information.

The persons or organisations who may be concerned in the preparation of the specification include:

> The Designer
> The Installer
> The Distributor of Electricity
> The Installation Owner and/or User
> The Architect
> Area Building Control bodies
> Any Regulatory Authority
> Any Licensing Authority
> The Health and Safety Executive

In producing the specification, advice should be sought from the installation owner and/or user as to the intended use. Often, as in a speculative building, the intended use is unknown. The specification and/or the operational manual must set out the basis of use for which the installation is suitable.

Precise details of each item of equipment should be obtained from the manufacturer   511.1
and/or supplier, and compliance with appropriate standards confirmed.

The operational manual must include a description of how the system as installed is to operate and all commissioning records. The manual should also include manufacturers' technical data for all items of switchgear, luminaires, accessories, etc. and any special instructions that may be needed. The Health and Safety at Work etc. Act 1974 Section 6 and the Construction (Design and Management) Regulations 2007 are concerned with the provision of information, and guidance on the preparation of technical manuals is given in the BS 4884 series (Technical manuals) and BS 4940 series (Technical information on construction products and services). The size and complexity of the installation will dictate the nature and extent of the manual.

General requirements

1

1.1 Safety

Electrical testing inherently involves some degree of hazard. It is the inspector's duty to ensure his (or her) own safety, and that of others, in the performance of the test procedures. The safety procedures detailed in Health and Safety Executive Guidance Note GS38 (revised) 'Electrical test equipment for use by electricians' should be observed. Where testing does not require the equipment or part of an installation to be live, it should be made dead and safely isolated. Guidance on live and dead working can be found in the Health and Safety Executive publication *Memorandum of Guidance on the Electricity at Work Regulations 1989* (HSR25). Guidance on safe isolation procedures can be found in a best practice guide published by the Electrical Safety Council.

When using test instruments, safety is best achieved by precautions such as:

1 understanding the equipment to be used and its rating
2 checking that all safety procedures are followed
3 checking that the instruments being used conform to the appropriate British Standard safety specifications. These are BS EN 61010 (Safety Requirements for Electrical Equipment for Measurement, Control, and Laboratory Use) and BS 5458:1977 (1993) Specification for Safety Requirements for Indicating and Recording Electrical Measuring Instruments and their Accessories. BS 5458 has been withdrawn, but is the standard to which older instruments should have been manufactured
4 checking that test leads including any probes or clips used are in good order, are clean and have no cracked or broken insulation. Where appropriate, the requirements of the Health and Safety Executive Guidance Note GS38 should be observed for test leads. This recommends the use of fused test leads aimed primarily at reducing the risks associated with arcing under fault conditions.

Particular attention should be paid to the safety aspects associated with any tests performed with instruments capable of generating a test voltage greater than 50 V, or which use the supply voltage for the purposes of the test in earth loop testing and residual current device (RCD) testing. Note the warnings given in Section 2.7.14 and Chapter 4 of this Guidance Note.

Electric shock hazards can arise from, for example, capacitive loads such as cables charged in the process of an insulation test, or voltages on the earthed metalwork whilst conducting a loop test or RCD test. The test limits quoted in these guidelines are intended to minimise the chances of receiving an electric shock during tests.

1.2   Required competence

610.1
610.5
611
The inspector carrying out the inspection and testing of any electrical installation must, as appropriate to his or her function, have a sound knowledge and experience relevant to the nature of the installation being inspected and tested, and to the technical standards. The inspector must also be fully versed in the inspection and testing procedures and employ suitable testing equipment during the inspection and testing process.

It is the responsibility of the inspector to:

610.1
612.1
1   ensure no danger occurs to any person or livestock and property is not damaged
2   compare the inspection and testing results with the design criteria
3   take a view on the condition of the installation and advise on remedial works

634.2   In the event of a dangerous situation being found, the inspector should recommend the immediate isolation of the defective part. The person ordering the work should be informed of this recommendation without delay.

1.3   The client

1.3.1   Certificates and Reports

631.1
632.1

631.2
634.1
Following the initial verification of a new installation or changes to an existing installation, an Electrical Installation Certificate, together with a schedule of inspections and a schedule of test results, is required to be given to the person ordering the work. In this context, 'work' means the installation work, not the work of carrying out the Inspection and Test. Likewise, following the periodic inspection and testing of an existing installation, a Periodic Inspection Report, together with a schedule of inspections and a schedule of test results, is required to be given to the person ordering the inspection.

Sometimes the person ordering the work is not the user. In such cases it is necessary for the user (e.g. employer or householder) to have a copy of the inspection and test documentation. It is recommended that those providing documentation to the person ordering the work recommend that the forms be passed to the user including any purchaser of a domestic property.

1.3.2   Rented domestic and residential accommodation

In England and Wales, The Landlord and Tenant Act 1985, Section 11 *Repairing obligations in short leases* sub-section (1)(b) requires a landlord 'to keep in repair and proper working order the installations in the dwelling-house for the supply of water, gas and electricity'.

A similar requirement can be found in the Housing (Scotland) Act 2006, Section 13 The repairing standard sub-section (1)(c) and Section 14 Landlord's duty to repair and maintain.

Although neither of the above Acts makes specific mention of a need to carry out periodic inspection and testing of the electrical installation, they do place an obligation on the landlord to maintain the installation in a condition suitable for the use intended.

Where an installation is not subject to an effective and ongoing planned and proactive maintenance programme, periodic inspection and testing carried out at appropriate intervals is a practical way of identifying where maintenance work is most required in order to keep the accommodation in a condition that is safe for use.

Any repairs must be carried out by a competent person. The Landlord is responsible for confirming the competency of any contractors carrying out such work.

1.4 Alterations and additions

Every alteration or addition to an existing installation must comply with the Regulations and must not impair the safety of the existing installation.

131.8
633.1
633.2

When inspecting and testing an alteration or addition to an electrical installation, the existing installation must be inspected and tested so far as is necessary to ensure the safety of the alteration or addition, including for example:

▶ protective conductor continuity
▶ earth fault loop impedance.

Whilst there is no obligation to inspect and test any part of the existing installation that does not affect and is not affected by the alteration or addition, observed departures are required to be noted in the comments box of electrical installation certificates (single signature or multiple signature) and minor works certificates.

633.2

1.5 Record keeping

It is a requirement that the appropriate documentation called for in 514.9, Part 6 and Part 7 is provided for every electrical installation.

132.13

Records of all checks, inspections and tests, including test results, should be kept throughout the working life of an electrical installation. This will enable deterioration to be identified. They can also be used as a management tool to ensure that maintenance checks are being carried out and to assess their effectiveness.

622
Appx 6 – guidance
to recipients

Regulation 20 (2) sub-section (e) of the Construction (Design and Management) Regulations 2007 requires a record known as 'the health and safety file' to be prepared, reviewed and updated. This should contain any information relating to the project which is likely to be needed during any subsequent construction work to ensure the health and safety of persons. Sub-section (f) requires that the health and safety file is passed on to the client on the completion of the construction work.

Regulation 17 (3) (3) requires that reasonable steps are taken to ensure that once the construction phase has been completed the information in the health and safety file remains available for inspection by any person who might need it to comply with any relevant statutory provisions. It also requires that the file is revised and updated as often as may be appropriate to incorporate any relevant new information.

Electrical installation certificates, minor electrical works certificates and periodic inspection reports would constitute relevant information in relation to this requirement.

Initial verification

2

2.1 Purpose of initial verification

Initial verification, in the context of Regulation 610.1, is intended to confirm that the installation complies with the design and has been designed and constructed in accordance with BS 7671.

610.1
611.2
612.1

This Section makes recommendations for the initial inspection and testing of electrical installations.

Chapter 61 of BS 7671 states the requirements for 'INITIAL VERIFICATION'. As far as reasonably practicable, an inspection shall be carried out to verify:

611.2

1 all installed electrical equipment and material is of the correct type and complies with applicable British Standards or acceptable equivalents

611.2

2 all parts of the fixed installation are correctly selected and erected

3 no part of the fixed installation is visibly damaged or otherwise defective.

Inspections

611

Inspection is an important element of inspection and testing, and is described in Section 2.6 of this Guidance Note.

Tests

612

The tests are described in Sections 2.7 and 3.10 of this Guidance Note.

Results

631.1

The results of inspection and tests are to be recorded as appropriate. The *Memorandum of Guidance on the Electricity at Work Regulations 1989* (HSR25) recommends records of all maintenance including test results be kept throughout the life of an installation – see guidance on EAWR Regulation 4(2). This can enable the condition of equipment and the effectiveness of maintenance to be monitored.

Relevant criteria

612.1

The relevant criteria are, for the most part, the requirements of the Regulations for the particular inspection or test. The criteria are given in Chapters 2 and 3 of this Guidance Note.

There will be some instances where the designer has specified requirements which are particular to the installation concerned. For example, the intended impedances may be different from those in BS 7671. In this case, the inspector should either ask for the design criteria or forward the test results to the designer for verification with the intended design. In the absence of such data the inspector should apply the requirements set out in BS 7671.

Verification

The responsibility for comparing inspection and test results with relevant criteria, as required by Regulation 612.1, lies with the party responsible for inspecting and testing the installation. This party, which may be the person carrying out the inspection and testing, should sign the inspection and testing box of the Electrical Installation Certificate or the declaration box of the Minor Electrical Installation Works Certificate. If the person carrying out the inspection and testing has also been responsible for the design and construction of the installation, he (or she) must also sign the design and construction boxes of the Electrical Installation Certificate, or make use of the single signature Electrical Installation Certificate.

2.2 Certificates

Appendix 6 of BS 7671 contains three model forms for the initial certification of a new installation or for an alteration or addition to an existing installation as follows:

▶ multiple signature Electrical Installation Certificate
▶ single signature Electrical Installation Certificate
▶ Minor Electrical Installation Works Certificate.

Examples of typical forms are given in Chapter 5.

Multiple signature Electrical Installation Certificate
The multiple signature certificate allows different persons to sign for design, construction, inspection and testing, and allows two signatories for design where there is mutual responsibility. Where designers are responsible for identifiably separate parts of an installation, the use of separate forms would be appropriate.

Single signature Electrical Installation Certificate
Where design, construction, inspection and testing are the responsibility of one person, a certificate with a single signature may replace the multiple signature form.

(See the 'short form' in Chapter 5.)

Minor Electrical Installation Works Certificate
This certificate is to be used only for minor works that do not include the provision of a new circuit, such as an additional socket-outlet or lighting point to an existing circuit.

The certificate may also be used for the replacement of equipment such as accessories or luminaires, but not for the replacement of distribution boards or similar items.

2.3 Required information

610.2 BS 7671 requires that the following information shall be made available to the person or persons carrying out the inspection and testing:

Assessment of general characteristics
311.1 **1** the maximum demand, expressed in amperes, kW or kVA per phase (after diversity is taken into account)
312.2 **2** the number and type of live conductors of the source(s) of energy and of the circuits used in the installation
312.3 **3** the type of earthing arrangement used by the installation and any facilities provided by the distributor for the user.

4 the nominal voltage(s) and its characteristics including harmonic distortion

313.1

5 the nature of the load current and supply frequency

6 the prospective fault current at the origin of the installation

7 the earth fault loop impedance (Z_e) of that part of the system external to the installation

8 the suitability for the requirements of the installation, including the maximum demand

9 the type and rating of the overcurrent protective device acting at the origin of the installation.

Note: These characteristics should also be available for safety services such as UPS and generators.

Diagrams, charts or tables

The Health and Safety at Work etc Act 1974 generally requires relevant information to be available as an aid to safe use, inspection, testing and maintenance. This may include those items listed in 514.9.1 shown below:

514.9.1

10 the type and composition of circuits, including points of utilisation, number and size of conductors and type of cable. This should include the Installation Method shown in Appendix 4 (paragraph 7) of BS 7671

11 the method used for compliance with the requirements for basic and fault protection and, where appropriate, the conditions required for automatic disconnection

410.3.2

12 the information necessary for the identification of each device performing the functions of protection, isolation and switching, and its location

13 any circuit or equipment vulnerable to a particular test.

2.4 Scope

It is essential that the inspector and the person ordering the inspection know the extent of the installation to be inspected and any criteria regarding the limit of the inspection. Both should be agreed and recorded on the Certificate.

2.5 Frequency of subsequent inspections

The proposed interval between periodic inspections is an element of the design, selection and erection of the installation. This interval is required to be noted on the Electrical Installation Certificate and on a notice to be fixed in a prominent position at or near the origin of the installation.

514.12.1

The person carrying out the inspection and testing should, in the light of their knowledge of the installation, its intended use and environmental factors, recommend when the first periodic inspection will be required. The information and tables in Chapter 3 of this Guidance Note have been prepared to provide guidance.

135.1
622.1

2.6 Initial inspection

2.6.1 General procedure

Inspection and, where appropriate, testing should be carried out and recorded on suitable schedules progressively throughout the different stages of erection and before the installation is certified and put into service.

A model Schedule of Inspections is shown in Chapter 5.

2.6.2 Comments on individual items to be inspected

The inspection should include at least the checking of those items listed in Section 611.3 of BS 7671.

526   **a   Connection of conductors**

Every connection between conductors and equipment/other conductors should
513.1   provide durable electrical continuity and adequate mechanical strength. Requirements
for the enclosure of and accessibility of connections must be considered.

b   Identification of cables and conductors and cords

514   Table 51 of BS 7671 provides a schedule of colour identification of each core of cables,
cords, bare conductors and flexible conductors.

514.3.3   It should be checked that each core or bare conductor is identified as necessary.
Busbar and pole colour should also comply with Table 51.

514.4.5   Where it is desired to indicate phase rotation, or a different function for cables of the
same colour, numbered or lettered sleeves are permitted. The single colour green
514.5.2   must not be used. The colour combination green and yellow is only to be used for
protective conductors.

c   Routing of cables

522   Cables should be routed as appropriate out of harm's way, and where necessary,
additionally protected against mechanical damage.

522.6.7   The 17th Edition has introduced a new requirement for cables installed in a wall or
partition at a depth of less than 50 mm from the surface where the installation is not
intended to be under the supervision of a skilled or instructed person. If the cable used
does not incorporate an earthed metallic covering; or, is not installed in an earthed
conduit, trunking or duct; or, is not provided with mechanical protection sufficient to
prevent damage being caused by nails, screws or similar, it will be necessary to provide
415.1.1   additional protection by means of an RCD having a rated residual current not exceeding
30 mA. This RCD should have an operating time not exceeding 40 ms at a residual
current of $5I_{\Delta n}$. This is a requirement even when cables are run within the permitted
cable routes described in 522.6.6 (v).

522.6.8   Another new requirement relates to cables installed in walls or partitions, the construction
of which contains metallic component parts such as studs, frames or skins where the
installation is not intended to be under the supervision of a skilled or instructed person.
Irrespective of the depth at which the cable has been installed, either the cable has
an earthed metallic covering, is run in an earthed wiring system, or given adequate
protection from penetration by nails, screws or similar, or additional protection by
means of RCD as described in the paragraph above will be required.

d   Cable selection

132.7   Where practicable, the cable size should be assessed against the protective arrangement
523   based upon information provided by the installation designer (where available).
524
525
Reference should be made, as appropriate, to Appendix 4 of BS 7671.

132.14.1   **e   Verification of polarity – single-pole device in a TN or TT system**

530.3.2   It must be verified that single-pole devices for protection or switching are installed in
line conductors only.

f   Accessories and equipment

553

Correct connection (suitability, polarity, etc.) must be checked.

Table 55.1 of BS 7671 is a schedule of types of plug and socket-outlet available, the ratings, and the associated British Standards.

Particular attention should be paid to the requirements for cable couplers.

553.2

Bayonet lampholders B15 and B22 should comply with BS EN 61184 and be of temperature rating T2.

559.6.1.7

g   Selection and erection to minimise the spread of fire

527

Fire barriers, suitable seals and/or protection against thermal effects should be provided if necessary to meet the requirements of BS 7671.

BS 7671 requires that each sealing arrangement be inspected to verify that it conforms to the manufacturer's erection instructions. This may be impossible without dismantling the system and it is essential, therefore, that inspection should be carried out at the appropriate stage of the work, and that this is recorded at the time for incorporation in the inspection and test documents.

527.2

h   Protection by provision of both basic and fault protection

This can be provided by SELV, PELV, double insulation or reinforced insulation.

412
414

For SELV and PELV, requirements include:

1   the nominal voltage must not exceed 50 V a.c. or 120 V d.c.
2   an isolated source, e.g. a safety isolating transformer to BS EN 61558-2-6
3   protective separation from all non SELV or PELV circuits
4   for SELV, basic insulation between the SELV system and earth
5   SELV exposed-conductive-parts must have no connection with earth, exposed-conductive-parts or protective conductors of other systems.

For double insulation, basic protection is provided by basic insulation, and fault protection is provided by supplementary insulation.

412.1.1

For reinforced insulation, both basic and fault protection is provided by reinforced insulation between live parts and accessible parts.

Where double or reinforced insulation is to be employed as the sole protective measure, it is important to confirm that the installation or circuit so protected remains under effective supervision to prevent any unauthorised change being made which could impair the effectiveness of the safety measure.

412.1.3

i   Basic protection

416

Insulation of live parts

Although protection by insulation is the usual method, there are other basic protection methods.

416.1

Barriers or enclosures

Where live parts are protected by barriers or enclosures, these should be checked for adequacy and security.

416.2

417.2 *Obstacles*

Protection by obstacles provides protection only against unintentional contact. If this method is used, the area shall be accessible only to skilled persons or to instructed persons under their supervision. This method of protection is not to be used in some installations and locations of increased shock risk. See Part 7 of BS 7671.

417.3 *Placing out of reach*

Placing out of reach also provides basic protection. Increased distances are necessary where long or bulky conducting objects are likely to be handled in the vicinity.

Bare live parts are only permitted in an area accessible to skilled persons, and the dimensions of passageways should be checked against the guidance in Appendix 3 of the *Memorandum of Guidance on the Electricity at Work Regulations 1989* (HSR25) issued by the Health and Safety Executive.

410.3.5 The protective measures of obstacles and placing out of reach may only be employed in installations or parts thereof where access is only possible for skilled or instructed persons under their supervision.

j   Fault protection

The methods of fault protection are classified in a number of sub-sections in BS 7671, and are:

411   **1**   automatic disconnection of supply

418.1   **2**   non-conducting location

418.2   **3**   earth-free local equipotential bonding

413   **4**   electrical separation.

418 Methods **2** and **3** are specialised protection systems. It is essential that inspection of such systems be carried out by persons competent in the discipline and having adequate information on the design of the system. For these specialised systems, the designer and client will advise of, and agree, the necessary effective and continuing supervision. This will also be the case where the protective measure of electrical separation is used to supply more than one item of current-using equipment.

411   **1**   *Automatic disconnection of supply (ADS)*

The presence, correct sizing, labelling and connection of appropriate protective conductors must be confirmed as follows:

542.3   ▶   earthing conductor

543   ▶   circuit protective conductors

544   ▶   protective bonding conductors

544.1   – main bonding conductors

544.2   – supplementary bonding conductors.

312.3 The earthing system must be determined, e.g.

   ▶   TN-C-S system (protective multiple earthing (PME) or protective neutral bonding (PNB))

   ▶   TN-S system

   ▶   TT system (earth electrode(s)).

411.4.5 The earth impedance must be appropriate for the protective device, i.e. RCD or overcurrent device.

Specialised systems

The specialised protective measures in Section 418 and discussed below may only be employed in installations or parts thereof which remain under the supervision of skilled persons at all material times, to prevent unauthorised changes being made which may render such protective methods ineffective.

410.3.6

2 *Non-conducting location*

When protection by this method is employed (e.g. in electronic equipment test areas or similar) all installed electrical equipment should meet the requirements of Section 416 with regard to provisions for basic protection.

418.1

418.1.1

Further, the exposed-conductive-parts of the installation should be so arranged that it is not possible for persons to make simultaneous contact with either two exposed-conductive-parts, or an exposed-conductive-part and any extraneous-conductive-part under normal operating conditions, if these parts are liable to be at different potentials as a result of failure of the basic insulation of a live part.

418.1.2

A non-conducting location should contain no protective conductors.

418.1.3

3 *Earth-free local equipotential bonding*

The use of earth-free equipotential bonding is intended to prevent the appearance of a dangerous touch voltage.

418.2

When protection by this method is employed, all installed electrical equipment should meet the requirements of Section 416 with regard to provisions for basic protection.

418.2.1

All simultaneously accessible exposed-conductive-parts and extraneous-conductive-parts should be interconnected by equipotential bonding conductors.

418.2.2

Measures must be taken to ensure that the local equipotential bonding conductors are not connected to Earth either directly or unintentionally via the exposed and extraneous-conductive parts to which they are connected.

418.2.3

A warning notice complying with 514.13.2 must be fixed in a prominent position adjacent to every point of access to the location concerned. This method is sometimes combined with 'electrical separation'.

418.2.5

Inspection should verify that no item is earthed within the area and that no earthed services or conductors enter or traverse the area, including the floor and ceiling. Inspection should confirm whether or not this has been achieved.

4 *Electrical separation*

This method may be applied to the supply of an individual item of equipment, or for more than one item of equipment.

413
418.3

Electrical separation is a protective measure where basic protection is provided by insulation of live parts, or by barriers/enclosures in accordance with Section 416, and fault protection is provided by simple separation of the separated circuit from other circuits and from earth.

If it is intended to supply more than one item of equipment using electrical separation, it will be necessary to meet the requirements of Section 418.3.

The separated circuit should be protected from damage and insulation failure.

418.3.3

418.3.4 Any exposed-conductive-parts should be connected together by insulated, non-earthed equipotential bonding conductors which should not be connected to the protective conductor or exposed-conductive-parts of any other circuit or to any extraneous-conductive-parts.

418.3.5 Any socket-outlet should have a protective conductor contact which is connected to the protective bonding system described above.

418.3.6 All flexible cables should contain a protective conductor for use as an equipotential bonding conductor except where such a cable supplies only items of equipment having double or reinforced insulation.

418.3.7 If two faults affecting two exposed-conductive-parts occur where these are fed by conductors of different polarity, a protective device should disconnect the supply in accordance with the disconnection time given in Table 41.1.

418.3.8 The product of the nominal voltage (volts) and length (metres) of the wiring system should not exceed 100,000 Vm and the length of the wiring system should not exceed 500 m.

k Additional protection

415.1.1 It should be confirmed that an RCD selected to provide additional protection has a rated residual current ($I_{\Delta n}$) not exceeding 30 mA.

415.1.2 It should be confirmed that appropriate measures for basic and fault protection are in accordance with Sections 411 to 414, as an RCD should not be used as the sole means of protection.

415.2.1 Where supplementary bonding is provided it should encompass all simultaneously accessible exposed-conductive-parts, extraneous-conductive-parts and the protective conductors of all equipment in the location where this protective measure is being applied.

415.2.2 The effectiveness of supplementary equipotential bonding as provided may be verified where the resistance between simultaneously accessible exposed and extraneous-conductive parts fulfils the following condition:

$R \leq 50$ V/I_a for a.c. systems

$R \leq 120$ V/I_a for d.c. systems

where I_a is the operating current of the protective device in amps; for overcurrent devices, this is the 5 second operating current, and for RCDs, $I_{\Delta n}$.

l Prevention of mutual detrimental influence

The requirements are stated in Regulation 132.11 and in Section 515: Mutual Detrimental Influence. Another aspect which should be taken into consideration during inspection is Section 528: Proximity to other Services.

m Isolating and switching devices

537 BS EN 60947 1 (Specification for Low Voltage Switchgear and Controlgear – General Rules) defines standard utilisation categories which allow for conditions of service use and the switching duty to be expected.

All switch utilisation categories must be appropriate for the nature of the load – see Table 2.1. Checking of utilisation category may need to be carried out during construction, if the label is obscured during erection. Guidance Note 2: *Isolation & Switching* provides more comprehensive guidance on this subject and should be consulted and its contents taken into account.

▼ **Table 2.1**

Examples of utilisation categories for alternating current installations

Utilisation category		Typical applications
Frequent operation	Infrequent operation	
AC-20A	AC-20B	Connecting and disconnecting under no-load conditions
AC-21A	AC-21B	Switching of resistive loads including moderate overloads
AC-22A	AC-22B	Switching of mixed resistive and inductive loads, including moderate overloads
AC-23A	AC-23B	Switching of motor loads or other highly inductive loads

If switchgear to BS EN 60947-1 is suitable for isolation it will be marked with the symbol:

This may be endorsed with a symbol advising of function, e.g. for a switch disconnector:

A summary of the suitability or otherwise of protective, isolation and switching devices to be employed for one or more of the functions of isolation, emergency switching and functional switching is given in Table 53.2.

Table 53.2

An isolation exercise should be carried out to check that effective isolation can be achieved. This should include, where appropriate, locking-off and inspection or testing to verify that the circuit is dead and no other source of supply is present.

n Presence of undervoltage protective devices

132.8
445

Suitable precautions should be in place where a reduction in voltage, or loss and subsequent restoration of voltage, could cause danger. Normally such a requirement concerns only motor circuits; if it is required it will have been specified by the designer.

o Protective devices

411
43

The choice and setting of each protective device including monitoring devices should be compared with the design.

514.8 **p Labelling of protective devices, switches and terminals**

A protective device must be arranged and identified so that the circuit protected may be easily recognised.

132.5
512.2 **q Selection of equipment and protective measures appropriate to external influences**
522
Equipment must be selected with regard to its suitability for the environment — ambient temperature, heat, water, foreign bodies, corrosion, impact, vibration, flora, fauna, radiation, building use and structure.

131.12 **r Adequacy of access to switchgear and equipment**
513
Every piece of equipment which requires operation or attention by a person must be so installed that adequate and safe means of access and sufficient working space are afforded.

514 **s Presence of danger notices and other warning notices**

Suitable warning notices, suitably located, are required to be installed to give warning of:

514.10 *Voltage*
- where a nominal voltage exceeding 230 V exists within an item of equipment or enclosure and where the presence of such a voltage would not normally be expected
- where a nominal voltage exceeding 230 V exists between simultaneously accessible terminals or other fixed live parts
- where different nominal voltages exist.

514.11 *Isolation*
- where live parts are not capable of being isolated by a single device. The location of disconnectors should also be indicated except where there is no possibility of confusion.

514.12.1 *Periodic inspection and testing*
- the wording of the notice is given in Regulation 514.12.1.

514.12.2 *RCDs*
- the wording of the notice is given in Regulation 514.12.2.

514.13 *Earthing and bonding connections*
514.13.1
- the requirements for the label and its wording are given in Regulation 514.13.1
- the wording of the notice required when protection by earth-free local
418.2.5
equipotential bonding (418.2.5 refers) or by electrical separation for the supply to more than one item of equipment (418.3 refers) is given in Regulation
514.13.2
514.13.2.

514.9 **t Presence of diagrams, instructions and similar information**

A schedule within or adjacent to the distribution board is sufficient for simple installations.

Part 5 **u Erection methods**

Chapter 52 contains detailed requirements on selection and erection. Fixings of switchgear, cables, conduit, fittings, etc. must be adequate for the environment.

2.6.3 Inspection checklist
(This checklist may also be used when carrying out periodic inspections.)

Listed below are requirements to be checked when carrying out an installation inspection. The list is not exhaustive.

General

1   Complies with requirements **1–3** in Section 2.1 (133.1, 134.1)
2   Accessible for operation, inspection and maintenance (513.1)
3   Suitable for local atmosphere and ambient temperature (Ch. 52). Installations in potentially explosive atmospheres are outside the scope of BS 7671. (See BS EN 60079 for electrical apparatus for use in an explosive gas atmosphere and BS EN 61241 for electrical apparatus for use in the presence of combustible dust.)
4   Circuits to be separate (no borrowed neutrals) (314.4)
5   Circuits to be identified (neutral and protective conductors in same sequence as line conductors) (514.1.2, 514.8.1)
6   Protective devices adequate for intended purpose (Ch. 53)
7   Disconnection times likely to be met by installed protective devices (Ch. 41)
8   Sufficient numbers of conveniently accessible socket-outlets are provided in accordance with the design (553.1.7). Note that in Scotland section 4.6.4 (socket-outlets) in the domestic technical handbook published by the Scottish Building Standards Agency (SBSA) gives specific recommendations for the number of socket-outlets for various locations within an installation.
9   All circuits suitably identified (514.1, 514.8, 514.9)
10 Suitable main switch provided (Ch. 53)
11 Supplies to any safety services suitably installed, e.g. Fire Alarms to BS 5839 and emergency lighting to BS 5266
12 Environmental IP requirements accounted for (BS EN 60529)
13 Means of isolation suitably labelled (514.1, 537.2.2.6)
14 Provision for disconnecting the neutral (537.2.1.7)
15 Main switches to single-phase installations, intended for use by an ordinary person, e.g. domestic, shop, office premises, to be double-pole (537.1.4)
16 RCDs provided where required (411.1, 411.3, 411.4, 411.5, 522.6.7, 522.6.8, 532.1, 701.411.3.3, 701.415.2, 702.55.4, 705.411.1, 705.422.7, 708.553.1.13, 709.531.2, 711.410.3.4, 711.411.3.3, 740.410.3, 753.415.1)
17 Discrimination between RCDs considered (314, 531.2.9)
18 Main earthing terminal provided (542.4.1) readily accessible and identified (514.13.1)
19 Provision for disconnecting earthing conductor (542.4.2)
20 Correct cable glands and gland plates used (BS 6121)
21 Cables used comply with British or Harmonized Standards (Appendix 4 of the Regulations, 521.1)
22 Conductors correctly identified (Section 514)
23 Earth tail pots installed where required on mineral insulated cables (134.1.4)
24 Non conductive finishes on enclosures removed to ensure good electrical connection and if necessary made good after connecting (526.1)
25 Adequately rated distribution boards (BS EN 60439 may require derating)
26 Correct fuses or circuit-breakers installed (Sections 531 and 533)
27 All connections secure (134.1.1)
28 Consideration paid to electromagnetic effects and electromechanical stresses (Ch. 52)
29 Overcurrent protection provided where applicable (Ch. 43)
30 Suitable segregation of circuits (Section 528)

31 Retest notice provided (514.12.1)

32 Sealing of the wiring system including fire barriers (527.2).

Switchgear

1 Suitable for the purpose intended (Ch. 53)

2 Meets requirements of BS EN 61008, BS EN 61009, BS EN 60947-2, BS EN 60898 or BS EN 60439 where applicable, or equivalent standards (511)

3 Securely fixed (134.1.1) and suitably labelled (514.1)

4 Non conductive finishes on switchgear removed at protective conductor connections and if necessary made good after connecting (526.1)

5 Suitable cable glands and gland plates used (526.1)

6 Correctly earthed (Ch. 54)

7 Conditions likely to be encountered taken account of, i.e. suitable for the foreseen environment (Section 522)

8 Where relevant correct IP rating applied (BS EN 60529)

9 Suitable as means of isolation, where applicable (537.2.2)

10 Complies with the requirements for locations containing a bath or shower (Section 701)

11 Need for isolation, mechanical maintenance, emergency and functional switching met (Section 537)

12 Firefighter's switch provided where required (537.6.1)

13 Switchgear suitably coloured where necessary (537.6.4)

14 All connections secure (Section 526)

15 Cables correctly terminated and identified (Sections 514 and 526)

16 No sharp edges on cable entries, screw heads, etc. which could cause damage to cables (522.8)

17 All covers and equipment in place and secure (Section 522.6.3)

18 Adequate access and working space (132.12 and Section 513).

Installation check lists

Wiring accessories

Reference should also be made to the recommendations contained in Approved Document M with regard to the heights at which socket-outlets, switches and other controls should be installed. See also the IEE publication *Electrician's Guide to the Building Regulations*.

General (applicable to each type of accessory)

1 Complies with BS 5733, BS 6220 or other appropriate standard (Section 511)

2 Box or other enclosure securely fixed (134. 1.1)

3 Metal box or other enclosure earthed (Ch. 54)

4 Edge of flush boxes not projecting above wall surface (134.1.1)

5 No sharp edges on cable entries, screw heads, etc. which could cause damage to cables (522.8)

6 Non sheathed cables, and cores of cable from which sheath has been removed, not exposed outside the enclosure (526.9)

7 Conductors correctly identified (514.6)

8 Bare protective conductors having a cross-sectional area of 6 mm² or less to be sleeved green and yellow (514.4.2, 543.3.2)

9 Terminals tight and containing all strands of the conductors (Section 526)

10 Cord grip correctly used or clips fitted to cables to prevent strain on the terminals (522.8.5, 526.6)

11 Adequate current rating (133.2.2)

12 Suitable for the conditions likely to be encountered (Section 522).

Lighting controls
1   Light switches comply with BS 3676 (Section 511)
2   Suitably located (Section 512.2)
3   Single-pole switches connected in line conductors only (132.14.1)
4   Correct colour coding or marking of conductors (514.6)
5   Earthing of exposed metalwork, e.g. metal switchplate (Ch. 54)
6   Complies with the requirements for locations containing a bath or shower (Section 701)
7   Adequate current rating (133.2.2)
8   Suitable for inductive circuits or derated where necessary (512.1.2)
9   Switch labelled to indicate purpose, where this is not obvious (514.1.1)
10 Appropriate controls suitable for the luminaires (559.6.1.9).

Lighting points
1   Correctly terminated in a suitable accessory or fitting (559.6.1.1)
2   Ceiling rose complies with BS 67 (559.6.1.1)
3   Not more than one flex unless designed for multiple pendants (559.6.1.3)
4   Flex support devices used (559.6.1.5)
5   Switch-lines identified (514.3.2)
6   Holes in ceiling above rose made good to prevent spread of fire (527.2.1)
7   Not connected to a supply exceeding 250 V (559.6.1.2)
8   Suitable for the mass suspended (559.6.1.5)
9   Lampholders to BS EN 60598 (559.6.1.1)
10 Luminaire couplers comply with BS 6972 or BS 7001 (559.6.1.1)
11 Track systems comply with BS EN 60570 (559.4.4).

Socket-outlets
1   Complies with BS 196, BS 546, BS 1363, BS EN 60309 2 (553.1.3) and shuttered for household and similar installations (553.1.4)
2   Mounting height above the floor or working surface suitable (553.1.6)
3   Correct polarity (612.6)
4   If installed in a location containing a bath or shower, installed beyond 3 m horizontally of the bath or shower unless shaver supply unit or SELV (701.512.3)
5   Protected where mounted in a floor (Section 522)
6   Not used to supply a water heater having uninsulated elements (554.3.3)
7   Circuit protective conductor connected directly to the earthing terminal of the socket outlet, on a sheathed wiring installation (543.2.7)
8   Earthing tail from the earthed metal box, on a conduit installation to the earthing terminal of the socket outlet (543.2.7).

Joint box
1   Joints accessible for inspection (526.3)
2   Joints protected against mechanical damage (526.7)
3   All conductors correctly connected (526.1).

Fused connection unit
1   Correct rating and fuse (533.1)
2   Complies with BS 1363-4 (559.6.1.1 vii).

Cooker control unit
1   Sited to one side and low enough for accessibility and to prevent flexes trailing across radiant plates (522.2.1)
2   Cable to cooker fixed to prevent strain on connections (522.8.5).

Conduits
General
1 Securely fixed, box lids in place and adequately protected against mechanical damage (522.8)
2 Inspection fittings accessible (522.8.6)
3 Number of cables for easy draw not exceeded (522.8.1 and see *On Site Guide* Appx 5)
4 Solid elbows and tees used only as permitted (522.8.1 and 522.8.3)
5 Ends of conduit reamed and bushed (522.8)
6 Adequate boxes suitably spaced (522.8 and see *On Site Guide* Appx 5)
7 Unused entries blanked off where necessary (412.2.2)
8 Conduit system components comply with a relevant British Standard (Section 511)
9 Provided with drainage holes and gaskets as necessary (522.3)
10 Radius of bends such that cables are not damaged (522.8.3)
11 Joints, scratches, etc. in metal conduit protected by painting (134.1.1, 522.5).

Rigid metal conduit
1 Complies with BS EN 50086 or BS EN 61386 (Section 511)
2 Connected to the main earth terminal (411.4.2)
3 Line and neutral cables contained in the same conduit (521.5.2)
4 Conduit suitable for damp and corrosive situations (522.3 and 522.5)
5 Maximum span between buildings without intermediate support (522.8 and see Guidance Note 1 and *On Site Guide* Appx 5).

Rigid non-metallic conduit
1 Complies with BS 4607, BS EN 60423, BS EN 50086-2-1 or the BS EN 61386 series (521.6)
2 Ambient and working temperatures within permitted limits (522.1 and 522.2)
3 Provision for expansion and contraction (522.8)
4 Boxes and fixings suitable for mass of luminaire suspended at expected temperature (522.8, 559.6.1.5).

Flexible metal conduit
1 Complies with BS EN 60423 and BS EN 50086-1 or the BS EN 61386 series (521.6)
2 Separate protective conductor provided (543.2.1)
3 Adequately supported and terminated (522.8).

Trunking
General
1 Complies with BS 4678 or BS EN 50085-1 (521.6)
2 Securely fixed and adequately protected against mechanical damage (522.8)
3 Selected, erected and routed so that no damage is caused by ingress of water (522.3)
4 Proximity to non-electrical services (528.2)
5 Internal sealing provided where necessary (527.2.4)
6 Holes surrounding trunking made good (527.2.1)
7 Band I circuits partitioned from Band II circuits or insulated for the highest voltage present (528.1)
8 Circuits partitioned from Band I circuits or wired in mineral-insulated metal-sheathed cables (528.1)
9 Common outlets for Band I and Band II provided with screens, barriers or partitions
10 Cables supported for vertical runs (522.8).

Metal trunking
1   Line and neutral cables contained in the same metal trunking (521.5.2)
2   Protected against damp or corrosion (522.3 and 522.5)
3   Earthed (411.4.2)
4   Joints mechanically sound and of adequate continuity (543.2.4).

Busbar trunking and powertrack systems
1   Busbar trunking to comply with BS EN 60439-2 or other appropriate standard and powertrack system to comply with BS EN 61534 series or other appropriate standard (521.4)
2   Securely fixed and adequately protected against mechanical damage (522.8)
3   Joints mechanically sound and of adequate continuity (543.2.4).

Insulated cables
Non-flexible cables
1   Correct type (521)
2   Correct current rating (523)
3   Protected against mechanical damage and abrasion (522.8)
4   Cables suitable for high or low ambient temperature as necessary (522.1)
5   Non sheathed cables protected by enclosure in conduit, duct or trunking (521.10)
6   Sheathed cables
    ▶ routed in allowed zones or mechanical protection provided (522.6.6)
    ▶ in the case of domestic or similar installations not under the supervision of skilled or instructed persons, additional protection is provided by RCD having $I_{\Delta n}$ not exceeding 30mA (522.6.7)
7   Cables in partitions containing metallic structural parts in domestic or similar installations not under the supervision of skilled or instructed persons should be
    ▶ provided with adequate mechanical protection to suit both the installation of the cable and its normal use
    ▶ provided with additional protection by RCD having $I_{\Delta n}$ not exceeding 30 mA (522.6.8)
8   Where exposed to direct sunlight, of a suitable type (522.11)
9   Not run in lift shaft unless part of the lift installation and of the permitted type (BS 5655 and BS EN 81-1) (528.3.5)
10 Buried cable correctly selected and installed for use (522.6.4)
11 Correctly selected and installed for use overhead (521)
12 Internal radii of bends not sufficiently tight as to cause damage to cables or to place undue stress on terminations to which they are connected (relevant BS, BS EN and 522.8.3)
13 Correctly supported (522.8.4 and 522.8.5)
14 Not exposed to water, etc. unless suitable for such exposure (522.3)
15 Metal sheaths and armour earthed (411.3.1.1)
16 Identified at terminations (514.3)
17 Joints and connections electrically and mechanically sound and adequately insulated (526.1 and 526.2)
18 All wires securely contained in terminals, etc. without strain (522.8.5 and Section 526)
19 Enclosure of terminals (Section 526)
20 Glands correctly selected and fitted with shrouds and supplementary earth tags as necessary (526.1)
21 Joints and connections mechanically sound and accessible for inspection, except as permitted otherwise (526.1 and 526.3).

Flexible cables and cords (521.9)
1. Correct type (521)
2. Correct current rating (Section 523)
3. Protected where exposed to mechanical damage (522.6 and 522.8)
4. Suitably sheathed where exposed to contact with water (522.3) and corrosive substances (522.5)
5. Protected where used for final connections to fixed apparatus, etc. (526.9)
6. Selected for resistance to damage by heat (522.1)
7. Segregation of Band I and Band II circuits (BS 6701 and Section 528)
8. Fire alarm and emergency lighting circuits segregated (BS 5839, BS 5266 and Section 528)
9. Cores correctly identified (514.3.2)
10. Joints to be made using appropriate means (526.2)
11. Where used as fixed wiring, relevant requirements met (521.9.3)
12. Final connections to portable equipment, a convenient length and connected as stated (553.1.7)
13. Final connections to other current-using equipment properly secured or arranged to prevent strain on connections (Section 526)
14. Mass supported by cable to not exceed values stated (559.11.6).

Protective conductors
1. Cables incorporating protective conductors comply with the relevant BS (Section 511)
2. Joints in metal conduit, duct or trunking comply with Regulations (543.3)
3. Flexible or pliable conduit to be supplemented by a protective conductor (543.2.1)
4. Minimum cross sectional area of copper conductors (543.1)
5. Copper conductors, other than strip, of 6 mm^2 or less protected by insulation (543.3.2)
6. Circuit protective conductor at termination of sheathed cables insulated with sleeving (543.3.2)
7. Bare circuit protective conductor protected against mechanical damage and corrosion (542.3 and 543.3.1)
8. Insulation, sleeving and terminations identified by colour combination green and yellow (514.3.1 and 514.4.2)
9. Joints electrically and mechanically sound (526.1)
10. Separate circuit protective conductors not less than 4 mm^2 if not protected against mechanical damage (543.1.1)
11. Main and supplementary bonding conductors of correct size (Section 544).

Enclosures
General
1. Suitable degree of protection (IP Code in BS EN 60529) appropriate to external influences (416.2, Section 522 and Part 7).

2.7 Initial testing

The test methods described in this section are the preferred test methods to be used; other appropriate test methods are not precluded.

2.7.1 Initial testing

The test results must be recorded on the Schedule(s) of Test Results and compared with relevant criteria. For example, relevant criteria for earth fault loop impedance may be provided by the designer, obtained as described in Section 2.7.14 or, where appropriate, obtained from Appendix B of this Guidance Note. 612.1

A model Schedule of Test Results is shown in Chapter 5.

2.7.2 Electrical Installation Certificate

Regulation 631.1 of BS 7671 requires that, upon completion of the verification of a new installation, or changes to an existing installation, an Electrical Installation Certificate based on the model given in Appendix 6 of BS 7671 shall be provided. Section 632 requires that: 631.1

1 the Electrical Installation Certificate be accompanied by a Schedule of Inspections and a Schedule of Test Results. These schedules shall be based upon the models given in Appendix 6 of BS 7671 632.1

2 the Schedule of Test Results shall identify every circuit, including its related protective device(s), and shall record the results of the appropriate tests and measurements detailed in Chapter 61 632.2

3 the Electrical Installation Certificate shall be compiled, signed/authenticated by a competent person or persons stating that to the best of their knowledge and belief the installation has been designed, constructed, inspected and tested in accordance with BS 7671, any permissible deviations being listed 631.4
632.3

4 any defects or omissions revealed by the Inspector shall be made good and inspected and tested again before the Electrical Installation Certificate is issued. 632.4

2.7.3 Model forms

Typical forms for use when carrying out inspection and testing are included in Chapter 5 of this Guidance Note.

2.7.4 The sequence of tests

Initial tests should be carried out in the following sequence where relevant:

a continuity of protective conductors, including main and supplementary bonding (2.7.5);

b continuity of ring final circuit conductors (2.7.6);

c insulation resistance (2.7.7);

d protection by SELV, PELV or by electrical separation (2.7.8);

e protection by barriers or enclosures provided during erection (2.7.9);

f insulation resistance of non-conducting floors and walls (2.7.10);

g polarity (2.7.11);

h earth electrode resistance (2.7.12);

i protection by automatic disconnection of the supply (2.7.13);

j earth fault loop impedance (2.7.14);

k additional protection (2.7.15);

l prospective fault current (2.7.16);

m phase sequence (2.7.17);

n functional testing (2.7.18);

o voltage drop (2.7.19).

2.7.5 Continuity of protective conductors including main and supplementary bonding

411.3.1.1   Regulation 411.3.1.1 requires that installations which provide protection against electric shock using automatic disconnection of supply must have a circuit protective conductor run to and terminated at each point in the wiring and at each accessory. An exception is made for a lampholder having no exposed-conductive-parts and suspended from such a point.

Test methods 1 and 2 are alternative ways of testing the continuity of protective conductors.

612.2.1   Every protective conductor, including circuit protective conductors, the earthing conductor and main and supplementary equipotential bonding conductors, should be tested to verify that the conductors are electrically sound and correctly connected.

Test method 1 detailed below, as well as checking the continuity of the protective conductor, also measures $(R_1 + R_2)$ which, when added to the external impedance (Z_e), enables the earth fault loop impedance (Z_S) to be checked against the design (see Section 2.7.14).

Note 1: $(R_1 + R_2)$ is the sum of the resistance of the line conductor R_1 and the circuit protective conductor R_2.

Note 2: The reading may be affected by parallel paths through exposed-conductive-parts and/or extraneous-conductive-parts.

It should also be recognised that test methods 1 and 2 can only be applied simply to an 'all insulated' installation. Installations incorporating steel conduit, steel trunking, micc and thermoplastic/swa (steel wire armour) cables will introduce parallel paths to protective conductors. Similarly, luminaires fitted in grid ceilings and suspended from steel structures in buildings may create parallel paths.

In such situations, unless a plug and socket arrangement has been incorporated in the lighting system by the designer, the $(R_1 + R_2)$ test will need to be carried out prior to fixing accessories and bonding straps to the metal enclosures and finally connecting protective conductors to luminaires. Under these circumstances some of the requirements may have to be visually inspected after the test has been completed. This consideration requires tests to be performed during the erection of an installation, in addition to tests at the completion stage.

Instrument – Use a low-resistance ohmmeter for these tests. Refer to Section 4.3.

The resistance readings obtained include the resistance of the test leads. *The resistance of the test leads should be measured and deducted from all resistance readings obtained unless the instrument can auto-null.*

Test method 1

Connect the line conductor to the protective conductor at the distribution board or consumer unit so as to include all the circuit. Then test between line and earth terminals at each outlet in the circuit. The measurement at the circuit's extremity should be recorded on the Schedule of Test Results as the value of $(R_1 + R_2)$ for the circuit under test.

See Figure 2.1a for test method connections.

REMEMBER TO REMOVE THE TEMPORARY LINK WHEN TESTING IS COMPLETE.

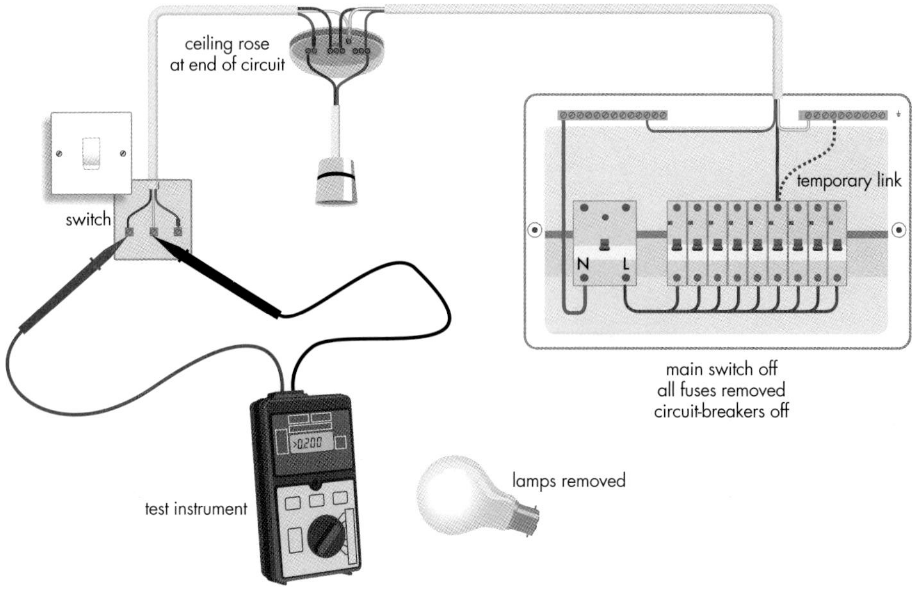

▼ **Figure 2.1a**
Connections for testing continuity of protective conductors: method 1

Test method 2

Connect one terminal of the continuity tester to the installation main earthing terminal, and with a test lead from the other terminal make contact with the protective conductors at various points on the circuit, e.g. luminaires, switches, spur outlets, etc. (see Figure 2.1b for test method connections). As at least one long lead is likely to be used when carrying out this test, it is essential to remember to either subtract the resistance of the lead from the overall measured resistance or, where such facility exists in the instrument, 'auto-null' the leads prior to carrying out the test.

This resistance R_2 is required to be recorded on the Schedule of Test Results.

To confirm the continuity of a bonding conductor, the leads from the instrument are connected to each end of the conductor and a reading is taken. One end of the bonding conductor and any intermediate connections with services may need to be disconnected to avoid parallel paths.

This method can also be used to confirm a bonding connection between extraneous-conductive-parts where it is not possible to see a bonding connection, e.g. where bonding clamps have been 'built in'. The test would be done by connecting the leads of the instrument between any two points such as metallic pipes between which a bonding connection was required and looking for a low (minimal deflection) reading of the order of 0.05 Ω or less.

Where metallic enclosures have been used as the protective conductors, e.g. conduit, trunking, steel-wire armouring, etc. the following procedure should be followed:

1 inspect the enclosure along its length for soundness of construction
2 perform the standard ohmmeter test using the appropriate test method described above.

Instrument: Use a low-resistance ohmmeter for this test. Refer to Section 4.3.

▼ **Figure 2.1b**

Connections for testing
continuity of protective
conductors: method 2

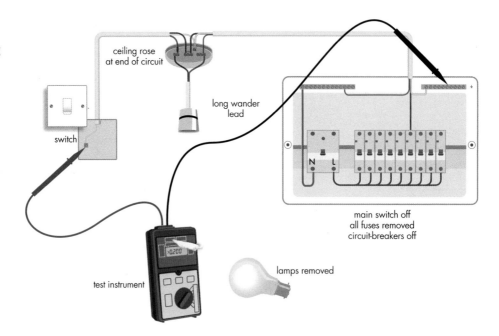

ceiling rose
at end of circuit

long wander
lead

switch

main switch off
all fuses removed
circuit-breakers off

lamps removed

test instrument

2.7.6 Continuity of ring final circuit conductors

A three-step test is required to verify the continuity of the line, neutral and protective conductors and correct wiring of every ring final circuit. The test results show if the ring has been interconnected to create an apparently continuous ring circuit which is in fact broken.

Instrument: Use a low-resistance ohmmeter for these tests. Refer to Section 4.3.

Step 1

The line, neutral and protective conductors are identified and the end-to-end resistance of each is measured separately (see Figure 2.2a). These resistances are r_1, r_n and r_2 respectively. A finite reading confirms that there is no open circuit on the ring conductors under test. The resistance values obtained should be the same (within 0.05 Ω) if the conductors are the same size. If the protective conductor has a reduced cross-sectional area, the resistance r_2 of the protective conductor loop will be proportionally higher than that of the line or neutral loop, e.g. 1.67 times for 2.5/1.5 mm^2 cable. If these relationships are not achieved then either the conductors are incorrectly identified or there is something wrong at one or more of the accessories.

Step 2

The line and neutral conductors are then connected together so that the outgoing line conductor is connected to the returning neutral conductor and vice versa (see Figure 2.2b). The resistance between line and neutral conductors is measured at each socket-outlet. The readings at each of the sockets wired into the ring will be substantially the same and the value will be approximately one-quarter of the resistance of the line plus the neutral loop resistances, i.e. $(r_1 + r_n)/4$. Any sockets wired as spurs will give a higher resistance value due to the resistance of the spur conductors.

Note: Where single-core cables are used, care should be taken to verify that the line and neutral conductors of opposite ends of the ring circuit are connected together. An error in this respect will be apparent from the readings taken at the socket-outlets, progressively increasing in value as readings are taken towards the midpoint of the ring, then decreasing again towards the other end of the ring.

Step 3

The above step is then repeated but with the line and cpc cross-connected (see Figure 2.2c). The resistance between line and earth is measured at each socket-outlet. The readings obtained at each of the sockets wired into the ring will be substantially the same and the value will be approximately one-quarter of the resistance of the line plus cpc loop resistances, i.e. $(r_1 + r_2)/4$. As before, a higher resistance value will be recorded at any sockets wired as spurs. The highest value recorded represents the maximum $(R_1 + R_2)$ of the circuit and is recorded on the Schedule of Test Results. The value can be used to determine the earth loop impedance (Z_S) of the circuit to verify compliance with the loop impedance requirements of the Regulations (see Section 2.7.14).

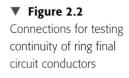

▼ **Figure 2.2**
Connections for testing continuity of ring final circuit conductors

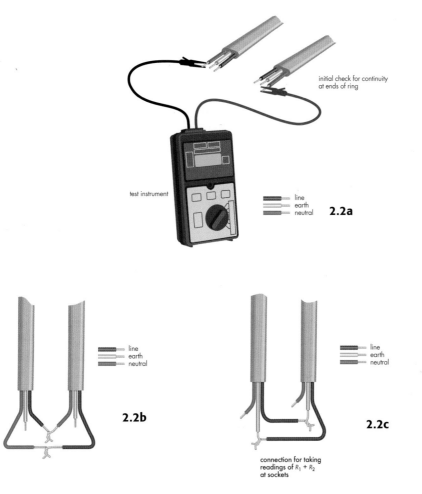

2.7.7 Insulation resistance

These tests are to verify that for compliance with BS 7671 the insulation of conductors and electrical accessories and equipment is satisfactory and that live conductors or protective conductors are not short-circuited, or do not show a low insulation resistance (which would indicate defective insulation).

612.3.2

Before testing check that:

1 pilot or indicator lamps, and capacitors are disconnected from circuits to avoid an inaccurate test value being obtained
2 if test 1 is to be used, voltage-sensitive electronic equipment such as dimmer switches, touch switches, delay timers, power controllers, electronic starters for fluorescent lamps, emergency lighting, RCDs, etc. are disconnected so that they are not subjected to the test voltage

3 there is no electrical connection between any line or neutral conductor and earth.

If circuits contain voltage-sensitive devices, test 2, which measures insulation resistance from protective earth to line and neutral connected together, may be appropriate if vulnerable equipment is not to be disconnected. Further precautions may also be necessary to avoid damage to some electronic devices, and it may be necessary to consult the manufacturer of the equipment to identify necessary precautions.

Instrument: Use an insulation resistance tester for these tests. Refer to Section 4.4.

Insulation resistance tests should be carried out using the appropriate d.c. test voltage specified in Table 61 of BS 7671. The installation will be deemed to conform with the Regulations in this respect if the main switchboard, and each distribution circuit tested separately with all its final circuits connected, but with current-using equipment disconnected, has an insulation resistance not less than that specified in Table 61, which is reproduced here as Table 2.2.

▼ **Table 2.2**

Minimum values of insulation resistance

Table 61

Circuit nominal voltage	Test voltage d.c. (V)	Minimum insulation resistance (MΩ)
SELV and PELV	250	0.5
Up to and including 500 V with the exception of SELV and PELV but including FELV	500	1.0
Above 500 V	1000	1.0

Simple installations that contain no distribution circuits should be tested as a whole.

The tests should be carried out with the main switch off, all fuses in place, switches and circuit-breakers closed, lamps removed, and fluorescent and discharge luminaires and other equipment disconnected. Where the removal of lamps and/or the disconnection of current-using equipment is impracticable, the local switches controlling such lamps and/or equipment should be open.

To perform the test in a complex installation it may need to be subdivided into its component parts.

Although an insulation resistance value of not less than 1.0 MΩ complies with the Regulations, where an insulation resistance of less than 2 MΩ is recorded the possibility of a latent defect exists. In these circumstances, each circuit should be tested separately. This will help identify (i) whether one particular circuit in the installation has a lower insulation resistance value, possibly indicating a latent defect that should be rectified, or (ii) whether the low insulation resistance represents, for example, the summation of individual circuit insulation resistance and as such may not be a cause for concern.

It should be noted that more stringent requirements for insulation resistance are applicable for the wiring of fire alarm systems in buildings. Clause 38.2 (*Inspection and testing of wiring*) of BS 5839-1:2002 states a minimum measured insulation resistance of 2 MΩ between conductors, between each conductor and earth, and between each conductor and any screen. It continues by stating that actual insulation resistance of fire alarm system wiring should be sufficiently high to avoid nuisance fault indications, as control and indicating equipment (CIE) may incorporate fault sensing for insulation resistance to earth on circuit wiring.

Test 1 – Insulation resistance between live conductors

Test between the live conductors at the appropriate distribution board.

Resistance readings obtained should be not less than the minimum values referred to in Table 2.2.

See Figure 2.3a for test method connections.

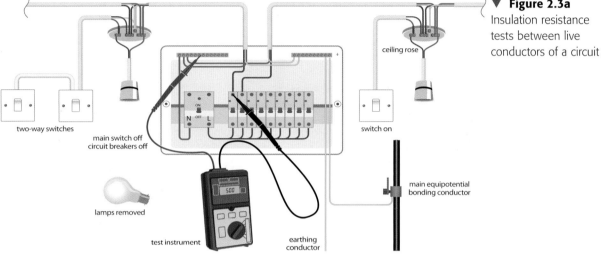

▼ **Figure 2.3a**
Insulation resistance tests between live conductors of a circuit

Note: The test will initially be carried out on the complete installation (see paragraph 2.7.7)

Test 2 – Insulation resistance to earth

Single-phase

Test between the line and neutral conductors connected together and earth at the appropriate distribution board, or for circuits/equipment not vulnerable to insulation resistance testing, line and neutral separately to earth.

For circuits containing two-way switching or two-way and intermediate switching, the switches must be operated one at a time and the circuits subjected to additional insulation resistance tests.

Three-phase

Test to earth from all live conductors (including the neutral) connected together, or for circuits/equipment not vulnerable to insulation resistance tests, each live conductor separately to earth.

Where a low reading is obtained (less than 2 MΩ) it may be necessary to test each conductor separately to earth, after ensuring that all equipment is disconnected.

Resistance readings obtained should be not less than the minimum values referred to in Table 2.2.

See Figure 2.3b for test method connections.

▼ **Figure 2.3b**
Insulation resistance
tests to earth

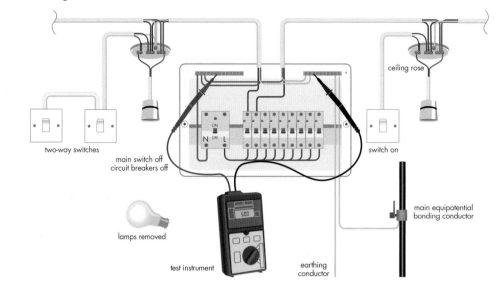

Note: The test will initially be carried out on the complete installation (see paragraph 2.7.7)

2.7.8 Protection by SELV, PELV or electrical separation

Where protection is provided by SELV, PELV or electrical separation, the measured insulation resistance should be at least that required by Table 2.2 for the circuit with the highest voltage present in the location. As such, if protection by SELV is being employed in an area which also contains circuits at low voltage, then the minimum acceptable value of insulation resistance for the SELV circuit wiring would be no lower than that obtained for the low voltage circuit and not less than 1 MΩ.

612.3.4

Table 61

SELV

The source of the SELV (separated extra-low voltage) supply should be inspected for conformity with Regulation 414.3, and if necessary the voltage should be measured to confirm that it does not exceed 50 V a.c. or 120 V d.c. If the voltage exceeds 25 V a.c. or 70 V d.c. (60 V d.c. ripple free), the means of basic protection should be verified. This may be by either:

414.1.1
414.4.5

416.2   **1**   inspection of the barriers or enclosures to ensure they offer a degree of protection not less than IP2X (IPXXB) or IP4X as required by Regulation 416.2, or

416.1   **2**   inspection of the insulation.

Table 61   The live conductors of any adjacent higher voltage circuit in contact with or in the same enclosure as SELV circuits must be tested to verify electrical separation in accordance with BS 7671.

This is achieved by testing between the live conductors of each SELV circuit connected together and the live conductors of any adjacent higher voltage circuits connected together.

Testing of SELV circuits

612.4.1   The first test applied to this arrangement is an insulation resistance test in accordance
Table 61   with Table 2.2.

Where the circuit is supplied from a safety source complying with BS EN 61558-2-6, the test is made at 250 V d.c.

414.3(i)

Instrument: Use an insulation resistance tester for this test. Refer to Section 4.4.

The SELV circuit conductors should be inspected to verify compliance with BS 7671.

414.4.1
414.4.2

Tests between SELV circuits and other circuits
To demonstrate compliance with BS 7671, an additional insulation resistance test should be applied at 500 V d.c. between the live conductors of the SELV circuit and those of other circuits.

Instrument: Use an insulation resistance tester for this test. Refer to Section 4.4.

Compliance with Regulation 414.4.4 will have been verified by the inspection required by Regulation 414.4.2. This should be confirmed.

414.4.4
414.4.2

The exposed-conductive-parts should be inspected to confirm compliance with Regulation 414.4.4.

414.4.4

PELV
PELV (protective extra-low voltage) installations are inspected and tested as for SELV installations except that an insulation test is not made between PELV circuits and Earth.

612.4.2
414.4.1

(PELV systems may include a protective conductor connected to the protective conductor of the primary circuit. Basic protection in the secondary circuit is dependent upon the primary circuit protection.)

Functional extra-low voltage
Extra-low voltage circuits not meeting the requirements for SELV or PELV are inspected and tested as low voltage circuits.

612.4.4

Electrical separation
The source of supply should be inspected to confirm compliance with the Regulations. In addition, should any doubt exist, the voltage should be measured to verify it does not exceed 500 V.

612.4.3

413.3.2

The insulation between live parts of the separated circuit and any other adjacent conductor (in the same enclosure or touching), and/or to earth, must be tested. This test should be performed at a voltage of 500 V d.c. and the insulation resistance should be not less than 1.0 MΩ.

612.4.3

Instrument: Use an insulation resistance tester for this test. Refer to Section 4.4.

The live parts of the separated circuit must be tested to ensure that they are electrically separate from other circuits. This is achieved by testing between the live conductors of the separated circuit connected together and the conductors of any other adjacent circuit strapped together.

413.3.2

The first test applied to this arrangement is an insulation resistance test at 500 V d.c. The insulation resistance should be not less than 1.0 MΩ.

Instrument: Use an insulation resistance tester for this test. Refer to Section 4.4.

413.3.5   A separate wiring system is preferred for electrical separation. If multicore cables or insulated cables in insulated conduit are used, all cables must be insulated to the highest voltage present and each circuit must be protected against overcurrent.

612.4.3
413.3.3   A 500 V d.c. insulation resistance test is performed between the exposed-conductive-parts of any item of connected equipment, and the protective conductor or exposed-conductive-parts of any other circuit, to confirm compliance with the Regulations. The insulation resistance should be not less than 1.0 MΩ.

Instrument: Use an insulation resistance tester for this test. Refer to Section 4.4.

418.3   If the separated circuit supplies more than one item of equipment, the requirements of the Regulations must be verified as follows.

1   Apply a continuity test between all exposed-conductive-parts of the separated circuit to ensure they are bonded together. This equipotential bonding is then subjected to a 500 V d.c. insulation resistance test between it and the protective conductor or exposed-conductive-parts of other circuits, or to extraneous-conductive-parts. The insulation resistance should be not less than 1.0 MΩ.
Instrument: Use an insulation resistance tester for this test. Refer to Section 4.4.

2   All socket-outlets must be inspected to ensure that the protective conductor contact is connected to the equipotential bonding conductor.

3   All flexible cables other than those feeding Class II equipment must be inspected to ensure that they contain a protective conductor for use as an equipotential bonding conductor.

Table 41.2
Table 41.3
Table 41.4   **4**   Operation of the protective device must be verified by measurement of the fault loop impedances (i.e. between live conductors) to the various items of connected equipment. These values can then be compared with the relevant maximum Z_s values given by Regulation 411.4.6 for fuses and 411.4.7 for circuit-breakers (Tables 41.2, 41.3 and 41.4 for 230 V systems), with reference to the type and rating of the protective device for the separated circuit. If protection is provided by overcurrent devices, the appropriate value of loop impedance given by Regulation 411.4.5 (Tables 41.2, 41.3 and 41.4 for 230 V systems) shall be determined. Although these tables pertain to the line/protective conductor loop path, and the measured values are between live conductors, they give a reasonable approximation to the values required to achieve the required disconnection time.
Instrument: Use an earth fault loop impedance tester for this test. Refer to Section 4.5.

2.7.9 Protection by barriers or enclosures provided during erection

612.4.5
416.2.1
416.2.2
416.2.3
416.2.4   This test is not applicable to barriers or enclosures of factory-built equipment. It is applicable to those provided on site during the course of assembly or erection and therefore is seldom necessary. Where, during erection, an enclosure or barrier is provided to provide basic protection, a degree of protection not less than IP2X or IPXXB is required. Readily accessible horizontal top surfaces must have a degree of protection of at least IP4X or IPXXD.

416.2.1   The degree of protection afforded by IP2X is defined in BS EN 60529 as protection against the entry of 'Fingers or similar objects not exceeding 80 mm in length. Solid objects exceeding 12 mm in diameter'. The test is made with a metallic standard test finger (test finger 1 to BS 61032).

Both joints of the finger may be bent through 90 ° with respect to the axis of the finger, but in one and the same direction only. The finger is pushed without undue force (not more than 10 N) against any openings in the enclosure and, if it enters, it is placed in every possible position.

A SELV supply, not exceeding 50 V, in series with a suitable lamp is connected between the test finger and the live parts inside the enclosure. Conducting parts covered only with varnish or paint, or protected by oxidation or by a similar process, must be covered with a metal foil electrically connected to those parts which are normally live in service.

The protection is satisfactory if the lamp does not light.

The degree of protection afforded by IP4X is defined in BS EN 60529 as protection against the entry of 'Wires or strips of thickness greater than 1.0 mm, and solid objects exceeding 1.0 mm in diameter'.

416.2.2

The test is made with a straight rigid steel wire of 1 mm + 0.05/0 mm diameter applied with a force of 1 N ± 10 per cent. The end of the wire must be free from burrs, and at a right angle to its length.

The protection is satisfactory if the wire cannot enter the enclosure.

Reference should be made to the appropriate product standard or BS EN 60529 for a fuller description of the degrees of protection, details of the standard test finger and other aspects of the tests.

2.7.10 Insulation resistance/impedance of floors and walls (protection by a non-conducting location)

Where fault protection is provided by a non-conducting location, the following should be verified:

418.1.2

1 exposed-conductive-parts should be arranged so that under normal circumstances a person will not come into simultaneous contact with:
   ▶ two exposed-conductive-parts or
   ▶ an exposed-conductive-part and any extraneous-conductive-part
2 in a non-conducting location there must be no protective conductors
3 any socket-outlets installed in a non-conducting location must not incorporate an earthing contact
4 the resistance of insulating floors and walls to the main protective conductor of the installation should be tested at not less than three points on each relevant surface, one of which should be approximately 1 m from any extraneous-conductive-part, e.g. pipes, in the location. Methods of measuring the insulation resistance/impedance of floors and walls are described later in this section.

418.1.3

418.1.5
612.5.1

If at any point the resistance is less than the specified value (50 kΩ where the voltage to earth does not exceed 500 V) the floors and walls are deemed to be extraneous-conductive-parts.

418.1.5

If any extraneous-conductive-part is insulated it should, when tested at 500 V d.c., have an insulation resistance not less than 1 MΩ, and be capable of withstanding at least 2 kV a.c. rms without breakdown, and should not pass a leakage current exceeding 1 mA in normal use.

418.1.4(iii)

612.5.2

Measuring insulation resistance of floors and walls

Appx 13 A magneto-ohmmeter or battery-powered insulation tester providing a no-load voltage of approximately 500 V (or 1000 V if the rated voltage of the installation exceeds 500 V) is used as a d.c. source.

The resistance is measured between the test electrode and the main protective conductor of the installation.

The test electrodes may be either of the following types. In case of dispute, the use of test electrode 1 is the reference method.

It is recommended that the test be made before the application of the surface treatment (varnishes, paints and similar products).

Test electrode 1
The test electrode (see Figure 2.4a) comprises a metallic tripod of which the parts resting on the floor form the points of an equilateral triangle. Each supporting part is provided with a flexible base ensuring, when loaded, close contact with the surface being tested over an area of approximately 900 mm^2 and having a combined resistance of less than 5000 Ω between the terminal and the conductive rubber pads.

Before measurements are made, the surface being tested is cleaned with a cleaning liquid. While measurements of the floors and walls are being made, a force of approximately 750 N (floors) or 250 N (walls), is applied to the tripod.

▼ **Figure 2.4a**
Test electrode 1

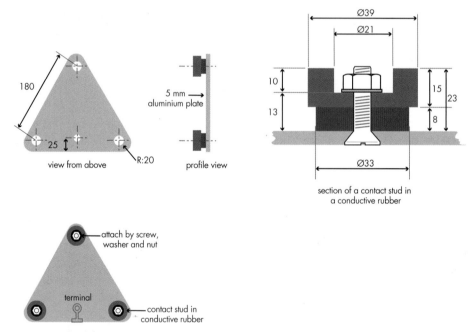

Test electrode 2
The electrode (see Figure 2.4b) comprises a square metallic plate with sides that measure 250 mm and a square of dampened water-absorbent paper or cloth, from which surplus water has been removed, with sides that measure 270 mm. The paper/cloth is placed between the metal plate and the surface being tested.

During the measurement a force of approximately 750 N (12 stone or 75 kg in weight) in the case of floors or 250 N in the case of walls is applied on the plate.

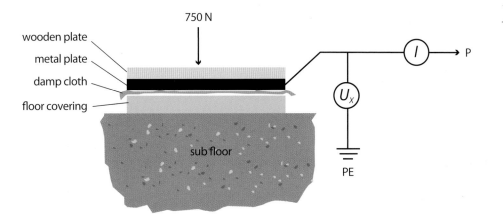

2.7.11 Polarity

The polarity of all circuits must be verified before connection to the supply, with either an ohmmeter or the continuity range of an insulation and continuity tester.

612.6

Instrument: Use a low-resistance ohmmeter for these tests. Refer to Section 4.3.

One method of test is as described for test method 1 (Section 2.7.5) for testing the continuity of protective conductors.

612.2.1

For radial circuits the $(R_1 + R_2)$ measurements, made as in test method 1 (Section 2.7.5), should be made at each point.

612.2.1

REMEMBER TO REMOVE THE TEMPORARY LINK WHEN TESTING IS COMPLETE.

For ring circuits, if the test required by Regulation 612.2.2 has been carried out (Section 2.7.6), the correct connections of line, neutral and circuit protective conductors will have been verified, subject to the following where applicable.

612.2.2

Different makes and types of accessories are not consistent in the relative position of the cable terminations to the socket tubes. Therefore, if the testing has been carried out with accessories in place, a visual inspection is required.

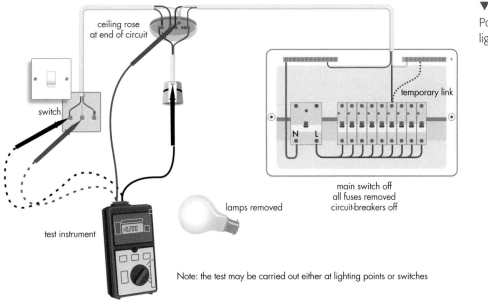

▼ **Figure 2.5**
Polarity test on a lighting circuit

It is necessary to check that all fuses and single-pole control and protective devices are connected in the line conductor. The centre contact of screw-type lampholders must be connected to the line conductor (except E14 & E27 to BS EN 60238). The line connection in socket-outlets and similar accessories must be connected to the line conductor.

See Figure 2.5 for test method connections.

After connection of the supply, polarity should be confirmed using an approved voltage indicator.

2.7.12   Earth electrode resistance
Measurement by standard method

612.7 When measuring earth electrode resistances to earth where low values are required,
542.1 as in the earthing of the neutral point of a transformer or generator, test method 1
542.2 below may be used.

Instrument: Use an earth electrode resistance tester for this test. Refer to Section 4.6.

Test method 1

Before this test is undertaken, the earthing conductor to the earth electrode must be disconnected either at the electrode or at the main earthing terminal to ensure that all the test current passes through the earth electrode alone. This will leave the installation unprotected against earth faults.

SWITCH OFF SUPPLY BEFORE DISCONNECTING THE EARTH.

542.2.2 The test should be carried out when the ground conditions are least favourable, such as during dry weather.

The test requires the use of two temporary test spikes (electrodes), and is carried out in the following manner.

Connection to the earth electrode is made using terminals C1 and P1 of a four-terminal earth tester. To exclude the resistance of the test leads from the resistance reading, individual leads should be taken from these terminals and connected separately to the electrode. If the test lead resistance is insignificant, the two terminals may be short-circuited at the tester and connection made with a single test lead, the same being true if using a three-terminal tester. Connection to the temporary spikes is made as shown in Figure 2.6.

The distance between the test spikes is important. If they are too close together, their resistance areas will overlap. In general, reliable results may be expected if the distance between the electrode under test and the current spike T1 is at least ten times the maximum dimension of the electrode system, e.g. 30 m for a 3 m long rod electrode.

With an auxiliary electrode T2 inserted halfway between the electrode under test E and temporary electrode T1, the voltage drop between E and T2 is measured. The resistance of the electrode is then obtained from the voltage between E and T2 divided by the current flowing between E and T1, provided that there is no overlap of the resistance areas.

To confirm that the electrode resistance obtained above is a true value, two further readings are taken, firstly with electrode T2 moved 6 m further from the electrode under test and secondly with electrode T2 moved 6 m closer to the electrode under test.

If the results obtained from the three tests above are substantially the same, the mean of the three readings is taken as the resistance of the earth electrode under test.

If the results obtained are significantly different, the above procedure should be repeated with test electrode T1 placed further from the electrode under test.

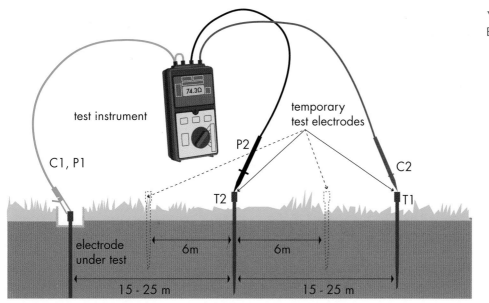

▼ **Figure 2.6**
Earth electrode test

The instrument output current may be a.c. or reversed d.c. to overcome electrolytic effects. Because these testers employ phase-sensitive detectors (psd), the errors associated with stray currents are eliminated.

The instrument should be capable of checking that the resistance of the temporary spikes used for testing is within the accuracy limits stated in the instrument specification. This may be achieved by an indicator provided on the instrument, or the instrument should have a sufficiently high upper range to enable a discrete test to be performed on the spikes.

If the temporary spike resistance is too high, measures to reduce the resistance will be necessary, such as driving the spikes deeper into the ground.

AFTER COMPLETION OF THE TESTING ENSURE THAT THE EARTHING CONDUCTOR IS RECONNECTED.

Earth electrode for RCDs

If the electrode under test is being used in conjunction with a residual current device the following method of test may be applied as an alternative to the earth electrode resistance test described above. In these circumstances, where the electrical resistances to earth are relatively high and precision is not required, an earth fault loop impedance tester may be used. Refer to Section 4.5.

Test method 2 (alternative for RCD protected TT installations)

Before this test is undertaken, the earthing conductor to the earth electrode should be disconnected at the main earthing terminal to ensure that all the test current passes through the earth electrode alone. This will leave the installation unprotected against earth faults.

SWITCH OFF SUPPLY BEFORE DISCONNECTING THE EARTH.

A loop impedance tester is connected between the line conductor at the source of the TT installation and the earth electrode, and a test performed. The impedance reading taken is treated as the electrode resistance.

BS 7671 requires

411.5.3 $$R_A I_{\Delta n} \leq 50 \text{ V}$$

where:

> R_A is the sum of the resistances of the earth electrode and the protective conductor(s) connecting it to the exposed-conductive-part
>
> $I_{\Delta n}$ is the rated residual operating current.

Table 41.5 Maximum values of R_A for the basic standard ratings of residual current devices are given in Table 2.3 for where 120 V $\leq U_0 \leq$ 230 V) unless the manufacturer declares alternative values.

▼ **Table 2.3**
Maximum values of earth electrode resistance for TT installations

RCD rated residual operating current $I_{\Delta n}$	Maximum value of earth electrode resistance, R_A
30 mA	1667 Ω
100 mA	500 Ω
300 mA	167 Ω
500 mA	100 Ω

The table indicates that the use of a suitably rated RCD will theoretically allow much higher values of R_A, and therefore of Z_s, than could be expected by using the overcurrent devices for fault protection. In practice, however, values above 200 Ω may not be stable, as soil conditions change due to factors such as soil drying and freezing.

AFTER THE TEST ENSURE THAT THE EARTHING CONDUCTOR IS RECONNECTED.

2.7.13 Protection by automatic disconnection of supply

The effectiveness of measures for fault protection by automatic disconnection of supply can be verified in installations having a TN system by:

▶ measurement of earth fault loop impedance (as described in 2.7.14 below)
▶ confirmation by visual inspection that overcurrent devices have suitable short-time or instantaneous tripping setting for circuit breakers, or current rating (I_n) and type for fuses
▶ where RCDs are employed, testing to confirm that the disconnection times of Chapter 41 of BS 7671 can be met.

In the case of installations supplied from a TT system, effectiveness can be verified by:

▶ measurement of the resistance of the earthing arrangement of the exposed-conductive-parts of the equipment for the circuit in question

▶ confirmation by visual inspection that overcurrent devices have suitable short-time or instantaneous tripping setting for circuit breakers, or current rating (I_n) and type for fuses

▶ where RCDs are employed, testing of the RCD to confirm that the disconnection times of Chapter 41 of BS 7671 can be met.

2.7.14 Earth fault loop impedance

Determining the earth fault loop impedance, Z_s

The earth fault current loop comprises the following elements, starting at the point of fault on the line–earth loop:

▶ the circuit protective conductor
▶ the main earthing terminal and earthing conductor
▶ for TN systems the metallic return path or, in the case of TT and IT systems, the earth return path
▶ the path through the earthed neutral point of the transformer
▶ the source line winding and
▶ the line conductor from the source to the point of fault.

Earth fault loop impedance Z_s may be determined by:

1 measurement of $R_1 + R_2$ during continuity testing (see Section 2.7.5) and adding to Z_e, i.e.

$$Z_s = Z_e + (R_1 + R_2)$$

612.9
612.8
Table 41.2
Table 41.3
Table 41.4

Z_e is determined by:

▶ measurement (see below), or
▶ enquiry of the electricity supplier, or
▶ calculation, or

2 direct measurement using an earth fault loop impedance tester (see below).

Determining Z_e

Measurement

The reasons that Z_e is required to be measured are twofold:

1 to verify that there is an earth connection
2 to verify that the Z_e value is equal to or less than the value determined by the designer and used in the design calculations.

Z_e is measured using an earth fault loop impedance tester at the origin of the installation.

542.4.2   The impedance measurement is made between the line of the supply and the means of earthing with the main switch open or with all the circuits isolated. The means of earthing must be disconnected from the installation earthed equipotential bonding for

610.1   the duration of the test to remove parallel paths. Care should be taken to avoid any shock hazard to the testing personnel and other persons on the site both whilst establishing contact, and performing the test. **ENSURE THAT THE EARTH CONNECTION HAS BEEN REPLACED BEFORE RECLOSING THE MAIN SWITCH.**

See Figure 2.7 for test method connections.

Instrument: Use an earth fault loop impedance tester for this test. Refer to Section 4.5.

▼ **Figure 2.7**
Test of Z_e at the origin of the installation

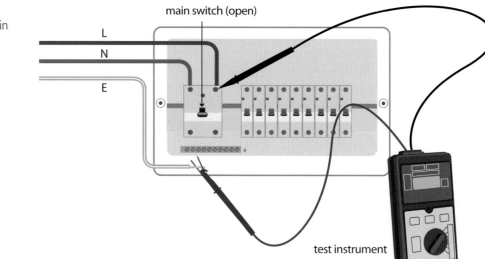

Measurement of $(R_1 + R_2)$ to add to Z_e

Whilst testing the continuity of protective conductors of radial circuits, or whilst testing the continuity of ring final circuits, the value of $(R_1 + R_2)_{test}$ is measured (at the test ambient temperature).

The measured value of $(R_1 + R_2)_{test}$ for the final circuit should be added to the value of $(R_1 + R_2)_{test}$ for any distribution circuit supplying the final circuit, to give the total $(R_1 + R_2)_{test}$ from the origin of the installation.

Direct measurement of Z_s

Direct measurement of Z_s can only be made on a live installation. Neither the connection with earth nor bonding conductors are disconnected. Readings given by the loop impedance tester may be less than $Z_e + (R_1 + R_2)$ because of parallel earth return paths provided by any bonded extraneous-conductive-parts. This must be taken into account when comparing the results with design data.

Care should be taken during the tests to avoid any shock hazard to the testing personnel, other persons or livestock on site.

Enquiry
Where Z_e is determined by enquiry of the electricity supplier, Z_s is then determined by adding $(R_1 + R_2)$ to this value of Z_e. However, a test must be made to ensure that the electricity supplier's earth terminal is actually connected with earth, using an earth fault loop impedance tester or a test lamp.

Verification of test results

Values of Z_s should be compared with one of the following:

612.1

1   for standard thermoplastic (PVC) circuits, the values in Appendix B of this Guidance Note
2   earth fault loop impedance figures provided by the designer
3   tabulated values in BS 7671, corrected for temperature
4   rule-of-thumb figures.

Table 41.2
Table 41.3
Table 41.4
Appx 14

1   *Standard thermoplastic (PVC) circuits*

The tabulated values given in Appendix B of this Guidance Note must not be exceeded when testing in an ambient temperature of 10 °C to 20 °C. As this is the normal temperature range to be expected, then correction for temperature is not usually required. Appendix B also provides a means of correcting these values for other test temperatures.

2   *Earth fault loop impedance figures provided by the designer*

This Guidance Note gives formulae which designers may use to obtain maximum values of Z_s for test purposes.

Examples of how these formulae may be used are given in Appendix B.

3   *Tabulated values in BS 7671, corrected for temperature*

The note below the tables in BS 7671 states that if the conductors are tested at a temperature which is different to their maximum permitted operating temperature, which they usually will be, then the reading should be adjusted accordingly.

Table 41.2
Table 41.3
Table 41.4

Appendix B provides a formula for making this adjustment, together with a worked example.

4   *Rule-of-thumb figures*

As a rule of thumb, the measured value of earth fault loop impedance for each circuit at the most remote outlet should not exceed 0.8 of the relevant value in the BS 7671 tables. This takes into account the increase of resistance of the conductors with the increase of temperature due to load current and errs on the side of safety.

Appx 14

Earth fault loop impedance test voltage

The normal method of test employed by a line–earth loop tester is to compare the unloaded loop circuit voltage with the circuit voltage when loaded with a low resistance, typically 10 Ω. This method of test can create an electric shock hazard if the line–earth loop impedance is high and the test duration is not limited. In these circumstances the potential of the protective conductor could approach line voltage for the duration of the test (see Section 4.5).

Residual current devices

The test (measuring) current of earth fault loop impedance testers may trip any RCD protecting the circuit. This will prevent a measurement being taken and may result in an unwanted disconnection of supply to the circuit under test.

Instrument manufacturers can supply loop testers that are less liable to trip RCDs. There are two common techniques:

1   testing at limited current
2   d.c. biasing the RCD.

1 *Limited current type*

Some instruments limit the test current to below 15 mA. This should mean that RCDs with a rated residual operating current of 30 mA and greater will not trip.

2 *d.c. biasing type*

Loop testers using a d.c. biasing technique saturate the core of the RCD prior to testing so that the test current is not detected. This technique can usually be expected to be effective for both type A and type AC RCDs.

Note: Type A RCDs – tripping is ensured for residual sinusoidal alternating currents and for pulsating direct currents.

Type AC RCDs – tripping is ensured for residual sinusoidal alternating currents.

2.7.15 Additional protection

415.1.1 Where an RCD with a rated residual operating current, $I_{\Delta n}$, not exceeding 30 mA is used to provide additional protection (against contact with live parts), the operating time of the device must not exceed 40 ms when subjected to a test current of 5 $I_{\Delta n}$. The maximum test time must not be longer than 40 ms, unless the protective conductor potential rises by less than 50 V. (The instrument supplier will advise on compliance.)

2.7.16 Prospective fault current

612.11 Regulation 612.11 requires that the prospective fault current, I_{pf}, under both short-circuit and earth fault conditions, be measured, calculated or determined by another method, at the origin and at other relevant points in the installation.

434.1 Regulation 612.11 introduces the requirements of Regulation 434.1 into the testing section. The designer is required to determine the prospective fault current, under both short-circuit and earth fault conditions, **at every relevant point of the installation**. This may be done by calculation, be ascertained by enquiry or be obtained directly using an instrument. The expression 'every relevant point' means every point where a protective device is required to operate under fault conditions, and includes the origin of the installation.

434.5.1 Regulation 434.5.1 states that the breaking capacity rating of each protective device shall be not less than the prospective fault current at its point of installation. The term prospective fault current includes the prospective short-circuit current and the prospective earth fault current. It is the greater of these two prospective fault currents which should be determined and compared with the breaking capacity of the device.

With the power on, the **maximum** value of the prospective short-circuit current can be obtained by direct connection of the instrument between live conductors at the protective device at the origin or other relevant location within the installation. Both two-lead and three-lead instruments capable of determining prospective fault current are available and it is important that any instrument being used is connected in accordance with the manufacturer's instructions for its use. Failure to do so could be dangerous, could result in damage to the instrument and may result in misleading readings being obtained.

With some instruments, the voltage between lines cannot be measured directly. Where this is the case, it can be assumed that for three-phase supplies, the maximum balanced prospective short-circuit level will be, as a rule of thumb, approximately twice the single-phase value. This figure errs on the side of safety.

Prospective earth fault current may be obtained with the same instrument. Again, care must be taken to ensure that the instrument has been connected as per manufacturer's recommendations.

The values obtained should be compared with the breaking capacity of the appropriate protective device. The breaking capacity of the protective device should be greater than the highest value of prospective fault current obtained using the instrument.

Whichever is the greater of the prospective short-circuit current and the prospective earth fault current obtained should be recorded on the Schedule of Test Results.

For a three-phase system, the prospective short-circuit current will always be larger than the earth fault current.

Instrument: Use the prospective fault current range of an earth fault loop impedance tester for this test. Refer to Section 4.5 – see final paragraph.

Rated short-circuit breaking capacities of protective devices
The rated short-circuit capacities of fuses, and circuit-breakers to BS 3871 (now withdrawn) and BS EN 60898 are shown in Table 2.4.

BS 3871 identified the short-circuit capacity of circuit-breakers with an 'M' rating.

Device type	Device designation	Rated short-circuit capacity (kA)	
Semi-enclosed fuse to BS 3036 with category of duty	S1A	1	
	S2A	2	
	S4A	4	
Cartridge fuse to BS 1361			
type 1		16.5	
type 2		33.0	
General purpose fuse to BS 88-2.1		50 at 415 V	
General purpose fuse to BS 88-6		16.5 at 240 V	
		80 at 415 V	
Circuit-breakers to BS 3871 (replaced by BS EN 60898)	M1	1	
	M1.5	1.5	
	M3	3	
	M4.5	4.5	
	M6	6	
	M9	9	
Circuit-breakers to BS EN 60898* and RCBOs to BS EN 61009*		I_{cn}	I_{cs}
		1.5	(1.5)
		3.0	(3.0)
		6	(6.0)
		10	(7.5)
		15	(7.5)
		20	(10.0)
		25	(12.5)

▼ **Table 2.4**
Rated short-circuit capacities

* Two rated short-circuit ratings are defined in BS EN 60898 and BS EN 61009:

I_{cn} the rated short-circuit capacity (marked on the device)
I_{cs} the service short-circuit capacity.

The difference between the two is the condition of the circuit-breaker after manufacturer's testing.

I_{cn} is the maximum fault current the breaker can interrupt safely, although the breaker may no longer be usable.

I_{cs} is the maximum fault current the breaker can interrupt safely without loss of performance.

The I_{cn} value is marked on the device in a rectangle e.g. $\boxed{6000}$ and for the majority of applications the prospective fault current at the terminals of the circuit-breaker should not exceed this value.

For domestic installations the prospective fault current is unlikely to exceed 6 kA, up to which value the I_{cn} will equal I_{cs}.

Where a service cut-out containing a cartridge fuse to BS 1361 type 2 supplies a consumer unit which complies with BS 5486-13 or BS EN 60439-3, then the short-circuit capacity of the overcurrent protective devices within consumer units may be taken to be 16 kA.

Fault currents up to 16 kA
Except for London and some other major city centres, the maximum fault current for 230 V single-phase supplies up to 100 A is unlikely to exceed 16 kA.

The short-circuit capacity of overcurrent protective devices incorporated within consumer units may be taken to be 16 kA where:

▶ the current ratings of the devices do not exceed 50 A
▶ the consumer unit complies with BS 5486-13 or BS EN 60439-3
▶ the consumer unit is supplied through a type 2 fuse to BS 1361:1971 rated at no more than 100 A.

Recording the prospective fault current
Both the Electrical Installation Certificate and the Periodic Inspection Report contain a box headed Nature of Supply Parameters, which requires the prospective fault current at the origin to be recorded. The value to be recorded is the greater of either the short-circuit current (between live conductors) or the earth fault current (between line conductor(s) and the main earthing terminal). If it is considered necessary to record values at other relevant points, they can be recorded on the Schedule of Test Results. Where the protective devices used at the origin have the necessary rated breaking capacity, and devices with similar breaking capacity are used throughout the installation, it can be assumed that the Regulations are satisfied in this respect for all distribution boards.

2.7.17 Phase sequence

612.12 The 17th Edition introduces in 612.12 a requirement for checking that phase sequence is maintained for multi-phase circuits within an installation. In practice, this will require confirmation that the phase rotation at three-phase distribution boards and the like is the same as that at the origin of the installation.

The main types of instrument suitable for carrying out such checks that are readily available are:

▶ rotating disc type
▶ indicator lamp type

Instruments containing both of the above forms of indication are also available.

Both types consist of an instrument body housing the indicator and three insulated and fused leads which are identified either by colour or as L1, L2 and L3. Typically at least one of the leads is fitted with an insulated 'crocodile' type clip to enable a single user to carry out the check.

In either case, the leads are connected in the correct sequence to the live conductors within the first piece of switchgear within the consumer's installation or as close to that point as is practically possible, and the phase sequence at that point is observed and noted.

In the case of a rotating disc type instrument, the disc will be rotating either clockwise or anticlockwise.

With the indicator lamp type either the L1/L2/L3 (formerly R/Y/B) lamp or the L1/L3/L2 (formerly R/B/Y) lamp will be illuminated.

The phase sequence is then checked at the other three-phase distribution boards, panels, etc. within the installation, and final connections corrected as necessary to maintain the correct sequence throughout the installation.

Both types of phase sequence indicator can also be used to verify phase sequence/ direction of rotation at the supply terminals to motors and to confirm the correct labelling/identification of plain conductors.

2.7.18 Functional testing
Operation of residual current devices
While the following tests are not a specific requirement of BS 7671, it is recommended that they are carried out.

In order to test the effectiveness of residual current devices after installation, a series of tests may be applied to check that they are within specification and operate satisfactorily. This test sequence will be in addition to proving that the test button is operational. The effectiveness of the test button should be checked after the test sequence. 612.8.1

For each of the tests, readings should be taken on both positive and negative half-cycles and the longer operating time recorded.

411.4.5 Prior to these RCD tests it is essential, for safety reasons, that the earth loop impedance is tested to check the requirements have been met.

Instrument: Use an RCD tester for these tests. Refer to Section 4.7.

Test method
The test is made on the load side of the RCD between the line conductor of the protected circuit and the associated cpc. The load should be disconnected during the test.

RCD testers require a few milliamperes to operate the instrument, and this is normally obtained from the line and neutral of the circuit under test. When testing a three-phase RCD protecting a three-wire circuit, the instrument's neutral is required to be connected to earth. This means that the test current will be increased by the instrument supply current and will cause some devices to operate during the 50 per cent test at a time when they should not operate. Under this circumstance it is necessary to check the operating parameters of the RCD with the manufacturer before failing the device.

610.1 These tests can result in a potentially dangerous voltage on exposed-conductive-parts and extraneous-conductive-parts when the earth fault loop impedance approaches the maximum acceptable limits. Precautions must therefore be taken to prevent contact of persons or livestock with such parts.

General purpose RCDs to BS 4293
1 with a leakage current flowing equivalent to 50 per cent of the rated tripping current of the RCD, the device should not open
2 with a leakage current flowing equivalent to 100 per cent of the rated tripping current of the RCD, the device should open in less than 200 ms.

Where the RCD incorporates an intentional time delay it should trip within a time range from 50 per cent of the rated time delay plus 200 ms to 100 per cent of the rated time delay plus 200 ms.

411.4.5 Because of the variability of the time delay it is not possible to specify a maximum test time. It is therefore imperative that the circuit protective conductor does not rise more than 50 V above earth potential ($Z_s\ I_{\Delta n} \leq 50$ V). It is suggested that in practice a 2 s maximum test time is sufficient.

General purpose RCDs to BS EN 61008 or RCBOs to BS EN 61009
1 with a leakage current flowing equivalent to 50 per cent of the rated tripping current of the RCD, the device should not open
2 with a leakage current flowing equivalent to 100 per cent of the rated tripping current of the RCD, the device should open in less than 300 ms unless it is of 'Type S' (or selective) which incorporates an intentional time delay, when it should trip within the time range from 130 ms to 500 ms.

RCD protected socket-outlets to BS 7288
1 with a test current flowing equivalent to 50 per cent of the rated tripping current of the RCD, the device should not open
2 with a test current flowing equivalent to 100 per cent of the rated tripping current of the RCD, the device should open in less than 200 ms.

Additional protection

Where an RCD with a rated residual operating current, $I_{\Delta n}$, not exceeding 30 mA is used to provide additional protection in the event of failure of basic protection and/ or the provision for fault protection or carelessness by users, the operating time of the device must not exceed 40 ms when subjected to a test current of $5I_{\Delta n}$. The maximum test time must not be longer than 40 ms, unless the protective conductor potential rises by less than 50 V. (The instrument supplier will advise on compliance.)

415.1
612.10

Integral test device

An integral test device is incorporated in each RCD. This device enables the functioning of the mechanical parts of the RCD to be verified by pressing the button marked 'T' or 'Test'.

Operation of the integral test device does not provide a means of checking:

1 the continuity of the earthing conductor or the associated circuit protective conductors, or
2 any earth electrode or other means of earthing, or
3 any other part of the associated installation earthing, or
4 the sensitivity of the device.

The RCD test button will only operate the RCD if it is energised.

Functional checks

All assemblies, including switchgear, controls and interlocks, are to be functionally tested, that is operated to confirm that they work and are properly installed, mounted and adjusted.

2.7.19 Verification of voltage drop

A new requirement has been introduced into BS 7671 that, where required it should be verified that voltage drop does not exceed the limits stated in relevant product standards of installed equipment. Where no such limits are stated, voltage drop should be such that it does not impair the proper and safe functioning of installed equipment.

612.14

Typically, voltage drop will be evaluated using the measured circuit impedance.

The requirements for voltage drop are deemed to be met where the voltage drop between the origin and the relevant piece of equipment does not exceed the values stated in Appendix 12.

Appendix 12 gives maximum values of voltage drop for either lighting or other uses depending upon whether the installation is supplied directly from an LV distribution system or from a private LV supply.

It should be remembered that voltage drop may exceed the values stated in Appendix 12 in situations such as motor starting periods and where equipment has a high inrush current where such events remain within the limits specified in the relevant product standard or reasonable recommendation by a manufacturer.

Verification of voltage drop is not normally required during initial verification.

Periodic inspection and testing 3

3.1 Purpose of periodic inspection and testing

The purpose of periodic inspection and testing is to provide, so far as is reasonably practicable, for:

621.2

1 the safety of persons and livestock against the effects of electric shock and burns in accordance with the general requirements of Regulation 131.1, and
2 protection against damage to property by fire and heat arising from an installation defect, and
3 confirmation that the installation is not damaged or deteriorated so as to impair safety, and
4 the identification of installation defects and non-compliance with the requirements of the Regulations which may give rise to danger.

For an installation under effective supervision in normal use, periodic inspection and testing may be replaced by an adequate regime of continuous monitoring and maintenance of the installation and all its constituent equipment by skilled persons. Appropriate maintenance records must be kept.

622.2

If an installation is maintained under a planned maintenance management system, incorporating monitoring and supervised by a suitably qualified electrical engineer, with the results being recorded and kept over a period of time, then a formal periodic inspection and test certificate may not be required. The records may be kept on paper or computer and should record that electrical maintenance and testing has been carried out. The results of any tests should be recorded. The results should be available for scrutiny and need not be in the standard IEE Periodic Inspection Report format.

3.2 Necessity for periodic inspection and testing

Periodic inspection and testing is necessary because all electrical installations deteriorate due to a number of factors such as damage, wear, tear, corrosion, excessive electrical loading, ageing and environmental influences. Consequently:

1 legislation requires that electrical installations are maintained in a safe condition and therefore must be periodically inspected and tested — Tables 3.1 and 3.2
2 licensing authorities, public bodies, insurance companies, mortgage lenders and others may require periodic inspection and testing of electrical installations, as is for example the case for houses in multiple occupation (HMOs) — Tables 3.1 and 3.2
3 additionally, periodic inspection and testing should be considered:
   a to assess compliance with BS 7671
   b on a change of occupancy of the premises
   c on a change of use of the premises

d after alterations or additions to the original installation

e because of any significant change in the electrical loading of the installation

f where there is reason to believe that damage may have been caused to the installation, as might be the case for example after flooding.

Reference to legislation and other documents is made below and it is vital that these requirements are ascertained before undertaking periodic inspection and testing.

3.3 Electricity at Work Regulations

Regulation 4(2) of the Electricity at Work Regulations 1989 requires that:

As may be necessary to prevent danger, all systems shall be maintained so as to prevent, so far as is reasonably practicable, such danger.

The *Memorandum of Guidance on the Electricity at Work Regulations 1989* (HSR25) published by the Health and Safety Executive advises that this regulation is concerned with the need for maintenance to ensure the safety of the system rather than being concerned with the activity of doing the maintenance in a safe manner, which is required by Regulation 4(3). The obligation to maintain a system arises if danger would otherwise result. There is no specific requirement to carry out a maintenance activity as such; what is required is that the system be kept in a safe condition. The frequency and nature of the maintenance must be such as to prevent danger so far as is reasonably practicable. Regular inspection of equipment including the electrical installation is an essential part of any preventive maintenance programme. This regular inspection may be carried out as required with or without dismantling and supplemented by testing.

There is no specific requirement to test the installation on every inspection. Where testing requires dismantling, the tester should consider whether the risks associated with dismantling and reassembling are justified. Dismantling, and particularly disconnection of cables or components, introduces a risk of unsatisfactory reassembly.

3.4 Design

When carrying out the design of an installation and particularly when specifying the equipment, the designer will be taking into account the quality of the maintenance to be specified including the frequency of routine checks and the period between subsequent inspections (supplemented as necessary by testing).

Information on the requirements for routine checks and inspections should be provided in accordance with Section 6 of the Health and Safety at Work etc. Act 1974 and as required by the Construction (Design and Management) Regulations 2007. Users of a premises should seek this information as the basis on which to make their own assessments. The Health and Safety Executive advise in their *Memorandum of Guidance on the Electricity at Work Regulations 1989* (HSR25), that practical experience of an installation's use may indicate the need for an adjustment to the frequency of checks and inspections. This is a matter of judgement for the duty holder. The Electrical Installation Certificate requires the designer's advice as to the intervals between inspections to be inserted on the certificate.

3.5 Routine checks

Electrical installations should not be left without any attention for the periods of years that are normally allowed between formal inspections. In domestic premises it is presumed that the occupier will soon notice any breakages or excessive wear and arrange for precautions to be taken and repairs to be carried out. In other situations, there must be arrangements made for initiating reports of wear and tear from users of the premises. This should be supplemented by routine checks. The frequency of these checks will depend entirely upon the nature of the premises. Routine checks would typically include:

▼ **Table 3.1**
Routine checks

Activity	Check
Defects reports	All reported defects have been rectified

Inspection	Look for:
	  breakages
	  wear/deterioration
	  signs of overheating
	  missing parts (covers, screws)
	  loose fixings
	
	Confirm:
	  switchgear accessible (not obstructed)
	  doors of enclosures secure
	  adequate labelling in place

Operation	Operate:
	  switchgear (where reasonable)
	  equipment – switch on and off
	  including RCDs (using test button)

These routine checks need not be carried out by an electrically skilled person but should be done by somebody who is able to safely use the installation and recognise defects.

3.6 Required information

It is essential that the inspector knows the extent of the installation to be inspected and any criteria regarding the limit of the inspection. This should be recorded.

Enquiries should be made to the person responsible for the electrical installation with regard to the provision of diagrams, design criteria, electricity supply and earthing arrangements.

514.9

Diagrams, charts or tables should be available to indicate the type and composition of circuits, identification of protective devices for shock protection, isolation and switching and a description of the method used for fault protection.

3.7 Frequency of inspection

The frequency of periodic inspection and testing must be determined taking into account:

622.1
1   the type of installation
2   its use and operation
3   the frequency and quality of maintenance
4   the external influences to which it is subjected.

Table 3.2 provides guidance on the initial frequency of formal inspections of electrical installations as well as the routine checks. The 'initial frequencies' in the title of the table refers to the time interval between the issuing of the Electrical Installation Certificate on completion of the work and the first inspection. However, the competent person carrying out subsequent inspections may recommend that the interval between future inspections be increased or decreased as a result of the findings of their inspection. For example, the inspector may recommend an interval greater than that suggested by Table 3.2 in instances where an installation has not suffered from damage or deterioration which would detract from its overall suitability for continued use. Conversely it would be appropriate to recommend that the next inspection is carried out sooner than is suggested in Table 3.2 where an installation has clearly not withstood the adverse effects of its environment and usage well and is not subject to adequate and appropriate maintenance.

In short, the inspector being a competent person should apply engineering judgement when deciding upon intervals between inspecting and testing an installation and may use the recommendations of Table 3.2 to suggest a suitable starting point for such a decision.

In the case of domestic and commercial premises, a change in occupancy of the premises may necessitate additional inspection and testing.

621.2
The formal inspections should be carried out in accordance with Chapter 62 of BS 7671. This requires an inspection comprising careful scrutiny of the installation, carried out without dismantling or with partial dismantling as required, together with the appropriate tests of Chapter 61.

Type of installation	Routine check sub-clause 3.5	Maximum period between inspections and testing as necessary	Reference (see key below)
1	**2**	**3**	**4**
General installation			
Domestic	–	Change of occupancy/10 years	
Commercial	1 year	Change of occupancy/5 years	1, 2
Educational establishments	4 months	5 years	1, 2
Hospitals	1 year	5 years	1, 2
Industrial	1 year	3 years	1, 2
Residential accommodation	at change of occupancy/1 year	5 years	1
Offices	1 year	5 years	1, 2
Shops	1 year	5 years	1, 2
Laboratories	1 year	5 years	1, 2
Buildings open to the public			
Cinemas	1 year	1 to 3 years	2, 6, 7
Church installations	1 year	5 years	2
Leisure complexes (excluding swimming pools)	1 year	3 years	1, 2, 6
Places of public entertainment	1 year	3 years	1, 2, 6
Restaurants and hotels	1 year	5 years	1, 2, 6
Theatres	1 year	3 years	2, 6, 7
Public houses	1 year	5 years	1, 2, 6
Village halls/community centres	1 year	5 years	1, 2
Special installations			
Agricultural and horticultural	1 year	3 years	1, 2
Caravans	1 year	3 years	8
Caravan parks	6 months	1 year	1, 2, 6
Highway power supplies	as convenient	6 years	
Marinas	4 months	1 year	1, 2
Fish farms	4 months	1 year	1, 2
Swimming pools	4 months	1 year	1, 2, 6
Emergency lighting	Daily/monthly	3 years	2, 3, 4
Fire alarms	Daily/weekly/monthly	1 year	2, 4, 5
Launderettes	1 year	1 year	1, 2, 6
Petrol filling stations	1 year	1 year	1, 2, 6
Construction site installations	3 months	3 months	1, 2

▼ **Table 3.2**
Recommended initial frequencies of inspection of electrical installations

Reference key

1   Particular attention must be taken to comply with SI 1988 No. 1057 – Electricity, Safety, Quality and Continuity Regulations 2002 (as amended).

2   SI 1989 No. 635 – Electricity at Work Regulations 1989 (Regulation 4 & *Memorandum*).

3   See BS 5266: Part 1: 2005 *Code of practice for the emergency lighting of premises other than cinemas and certain other specified premises used for entertainment.*

4   Other intervals are recommended for testing operation of batteries and generators.

5   See BS 5839: Part 1: 2002 *Code of practice for system design installation and servicing (Fire detection and alarm systems for buildings).*

6 Local Authority Conditions of Licence.

7 SI 1995 No. 1129 (Clause 27) – Cinematograph (Safety) Regulations.

8 It is recommended that a caravan is inspected and tested annually if it is used frequently (see 721.514.1 and Fig. 721 – Instructions for electricity supply).

3.8 Requirements for inspection and testing

3.8.1 General procedure

134
611.3(xvi)
514.9

Where diagrams, charts or tables are not available, a degree of exploratory work may be necessary so that inspection and testing can be carried out safely and effectively. A survey may be necessary to identify switchgear, controlgear and the circuits they control.

Note should be made of any known changes in environmental conditions, building structure, and alterations or additions which have affected the suitability of the wiring for its present load and method of installation.

610.1
621.3

During the inspection, the opportunity should be taken to identify dangers which might arise during the testing. Any location and equipment for which safety precautions may be necessary should be noted and the appropriate steps taken.

Periodic tests should be made in such a way as to minimise disturbance of the installation and inconvenience to the user. Where it is necessary to disconnect part or the whole of an installation in order to carry out a test, the disconnection should be made at a time agreed with the user and for the minimum period needed to carry out the test. Where more than one test necessitates a disconnection, where possible they should be made during one disconnection period.

612.3.2
612.3.3

A careful check should be made of the type of equipment on site so that the necessary precautions can be taken, where conditions require, to disconnect or short-out electronic and other equipment which may be damaged by testing. Special care must be taken where control and protective devices contain electronic components.

514.9

If the inspection and testing cannot be carried out safely without diagrams or equivalent information, Section 6 of the Health and Safety at Work etc. Act 1974 can be interpreted to require their preparation.

3.8.2 Sampling when carrying out inspection and testing

Considerable care and engineering judgement should be applied when deciding upon the extent of the installation that will be subjected to inspection and testing. The person carrying out the inspection/testing should obtain any relevant information with the assistance of the person ordering the work. This would include electrical installation certificates, minor works certificates, previous periodic inspection reports, maintenance records, site plans/drawings and data sheets relating to installed equipment. If the necessary information relating to the design, maintenance and modification of the installation cannot be obtained, it would be necessary to inspect/test a larger percentage, or in some circumstances 100 per cent, of the installation.

Other factors that would have a direct bearing upon the size of sample that would be acceptable would be:

- age and general condition of the electrical installation
- type and usage of an installation or part thereof
- ambient environmental conditions
- the effectiveness of ongoing maintenance
- period of time elapsed between previous inspections/tests
- the size of the installation.

Prior to starting the inspection/testing, the degree of sampling to be applied must be agreed with the person ordering the work. If significant defects or signs of damage or deterioration are identified within this agreed sample it is strongly recommended that the person ordering the work is approached and consent is obtained to carry out a more comprehensive inspection/test of the installation. If this larger, more comprehensive inspection/test yields further significant signs of deterioration, then representations should be made to inspect/test the whole of the installation.

It may be impractical to inspect/test the whole of a large installation at one time. If this is the case and sampling is agreed with the person ordering the work, it is important to examine different parts of the installation in subsequent inspections.

It would not be acceptable for the same area of the installation to be repeatedly inspected to the exclusion of other parts.

3.8.3 Scope
The requirement of BS 7671 for periodic inspection and testing is for INSPECTION 621.2 comprising a detailed examination of the installation without dismantling, or with partial dismantling as required, together with the tests of Chapter 61 considered appropriate by the person carrying out the inspection and testing. The scope of the periodic inspection and testing must be decided by a competent person, taking into account the availability of records and the use, condition and nature of the installation.

Consultation with the client or the client's representative prior to the periodic inspection and testing work being carried out is essential to determine the degree of disconnection which will be acceptable before planning the detailed inspection and testing.

For safety, it is necessary to carry out a visual inspection of the installation before testing or opening enclosures, removing covers, etc. So far as is reasonably practicable, the visual inspection must verify that the safety of persons, livestock and property is not endangered.

A thorough visual inspection should be made of all electrical equipment which is not concealed, and should include the accessible internal condition of a sample of the equipment. The external condition should be noted and if damage is identified or if the degree of protection has been impaired, this should be recorded on the Schedule of Inspections appended to the Report. The inspection should include a check on the condition of all electrical equipment and material, taking into account any available manufacturer's information, with regard to the following:

1 safety
2 wear and tear
3 corrosion
4 damage

5 excessive loading (overloading)
6 age
7 external influences
8 suitability

The assessment of condition should take account of known changes in conditions influencing and affecting electrical safety, e.g. extraneous-conductive-parts, plumbing, structural changes.

Where sections of an electrical installation are excluded from the scope of a Periodic Inspection and Test, they should be identified in the 'extent and limitations' box of the Report. However, such sections must not be permanently excluded from inspection and testing, and a suitable programme should be devised which includes the inspection and testing of such sections.

3.8.4 Isolation of supplies
The requirement of Regulation 14 of the Electricity at Work Regulations 1989 regarding working on or near live parts must be observed during inspection of an installation.

In domestic type premises the whole installation can be readily isolated for inspection, but with most other installations it is not practicable and too disruptive to isolate the whole installation for the amount of time that is required for a comprehensive inspection. Much of the inspection in such premises has to be done whilst the installation is in operation.

Main switch panels can rarely be isolated from the supply for long periods; similarly, the disruption that may be caused by isolating final circuit distribution boards for long periods often cannot be tolerated.

Distribution boards should be isolated separately for short periods for the internal inspection of live parts and examination of connections.

Where it is necessary to inspect live parts inside equipment, the supply to the equipment must be disconnected.

In order to minimise disruption to the operation of premises, the appropriate supplementary testing in Section 3.10 should be applied at the same time as the inspection.

3.9   Periodic inspection

3.9.1 Comments on individual items to be inspected
a   Joints and connections
611.3(i)
611.3(vi)
It is not practicable to inspect every joint and termination in an electrical installation. Nevertheless a sample inspection should be made. An inspection should be made of all accessible parts of the electrical installation, e.g. switchgear, distribution boards, and a sample of luminaire points and socket-outlets, to ensure that all terminal connections of the conductors are properly installed and secured. Any signs of overheating of conductors, terminations or equipment should be thoroughly investigated and included in the Report.

b   Conductors
The deterioration of, or damage to, conductors and their insulation, and their protective coverings, if any, should be noted.

The insulation and protective covering of each conductor at each distribution board of the electrical installation and at a sample of switchgear, luminaires, socket-outlets and other points, should be inspected to determine their condition and correct installation. There should be no signs of overheating, overloading or damage to the insulation, armour, sheath or conductors. 611.3(iv)

c Flexible cables and cords

Where a flexible cable or cord forms part of the fixed wiring installation, the inspection should include: 611.3(iii)
522

1 examination of the cable or cord for damage or defects
2 examination of the terminations and anchorages for damage or defects
3 the correctness of its installation with regard to additional mechanical protection, heat resistant sleeving, etc.

d Accessories and switchgear

It is recommended that a random sample of accessories and switchgear is given a thorough internal visual inspection of accessible parts to assess their electrical and mechanical condition. Where the inspection reveals: 611.3(v)
611.3(vi)

1 results significantly different from results recorded previously
2 results significantly different from results reasonably to be expected
3 adverse conditions, e.g. fluid ingress or worn or damaged mechanisms,

the inspection should be extended to include every switching device associated with the installation under inspection unless there is clear evidence of how the damage occurred.

Generally, it is not appropriate to apply sampling to socket-outlets and items of class I equipment.

e Protection against thermal effects

The presence of fire barriers, seals and means of protection against thermal effects should be verified, if reasonably practicable. 611.3vii)

f Basic and fault protection

611.3(viii)(a)

The protective measures SELV and PELV may be used to provide basic and fault protection. The requirements of 414 need to be checked, particularly with respect to the source, e.g. a safety isolating transformer to BS EN 61558-2-6, the need to separate the circuits, and the segregation of exposed-conductive-parts of the SELV system from any connection with the earthing of the primary circuits, or from any other connection with earth. 414
414.3
414.4.1
414.4.2
414.4.4

Under the requirements of the 17th Edition, the protective measures of double or reinforced insulation may also be used to provide both basic and fault protection, in which case the requirements of section 412 must be met. 412

g Basic protection

It should be established that the means of basic protection preventing access to live conductors is provided by one or more of the following methods: 611.3 (viii)(b)

1 insulation of live parts 612.3
2 installation of barriers or enclosures 612.4.5
3 obstacles
4 placing out of reach
5 SELV or PELV. 612.4.1
612.4.2

It should be established that the means of basic protection meets the requirements for the safety of any person or livestock from the effects of electric shock, fire and burns.

For each method for providing basic protection it should be established that there has been no deterioration or damage to insulation, no removal of barriers or obstacles and no alterations to enclosures or access to live conductors which would affect its effectiveness.

415.1.2   **IT SHOULD BE NOTED THAT AN RCD MUST NOT BE USED AS THE SOLE MEANS OF BASIC PROTECTION.**

h   Fault protection

612.3(viii)c
411

The method of fault protection must be determined and recorded. For automatic disconnection of supply, the adequacy of main equipotential bonding and the connection of all protective conductors with the earth is essential. The loss of such connections to the main earthing terminal may result in exposed and extraneous-conductive parts becoming hazardous-live in the event of a fault.

i   Protective devices

611.3(x)
611.3(xii)
132.12

The presence, accessibility, labelling and condition of devices for electrical protection, isolation and switching should be verified.

611.3(viii)c(i)

It should be established that each circuit is adequately protected with the correct type, size and rating of fuse or circuit-breaker. The suitability of each protective and monitoring device and its overload rating or setting should be checked.

It must be ascertained that each protective device is correctly located and appropriate to the type of earthing system and to the circuits protected.

611.3(xiv)
611.3(xii)

Each device for protection, isolation and switching should be readily accessible for normal operation, maintenance and inspection, and be suitably labelled where necessary.

611.3(x)

It should be established that a means of emergency switching, or where appropriate emergency stopping, is provided where required by the Regulations, so that the supply can be cut off rapidly to prevent or remove danger. Where a risk of electric shock is involved then the emergency switching device should switch all the live conductors except as permitted by the Regulations.

When conditions permit, an internal inspection should be made of any emergency switching device and tests should be carried out as described in Section 3.10.3.

j   Enclosures and mechanical protection

612.4.5
132.5
611.3(xiii)
416.2.4

The enclosure and mechanical protection of all electrical equipment should be inspected to ensure that they remain adequate for the type of protection intended. All secondary barriers (to IP2X or IPXXB) should be in place.

k   Marking and labelling

514.1.1

The labelling of each circuit should be verified.

514.1.2
514.1.3

It should be established that adjacent to every fuse or circuit-breaker there is a label correctly indicating the size and type of the fuse, nominal current of the circuit-breaker and identification of the protected circuit.

It should be ascertained that all switching devices are correctly labelled and identify the circuits controlled.

Notices or labels are required at the following points and equipment within an installation.

1 At the origin of every installation:

514.12.1

a notice of such durable material as to be likely to remain easily legible throughout the life of the installation, marked in indelible characters no smaller than the example in BS 7671 should be fixed in a prominent position at or near the origin of the installation.

IMPORTANT

This installation should be periodically inspected and tested and a report on its condition obtained, as prescribed in the IEE Wiring Regulations BS 7671 Requirements for Electrical Installations.

Date of last inspection...

Recommended date of next inspection............................

---

2 Where different voltages are present:

514.10

a equipment or enclosures within which a nominal voltage exceeding 230 V exists and where the presence of such a voltage would not normally be expected

b where terminals or other fixed live parts between which a nominal voltage exceeding 230 V exists are housed in separate enclosures or items of equipment which, although separated, can be reached simultaneously by a person

c means of access to all live parts of switchgear and other fixed live parts where different nominal voltages exist.

3 At earthing and bonding connections:

a a permanent label to BS 951 with the words

514.13.1

Safety Electrical Connection – Do Not Remove

---

should be permanently fixed in a visible position at or near:

1 the point of connection of every earthing conductor to an earth electrode, and

2 the point of connection of every bonding conductor to an extraneous-conductive-part, and

3 the main earth terminal, where separate from main switchgear.

514.13.2 **b** In an *earth-free* situation all exposed metalwork must be bonded together but **not** to the main earthing system. In this situation a suitable permanent notice durably marked in legible type and no smaller than the example in BS 7671 should be permanently fixed in a visible position.

> The protective bonding conductors associated with the electrical installation in this location **MUST NOT BE CONNECTED TO EARTH**. Equipment having exposed-conductive-parts connected to Earth must not be brought into this location.

514.12.2 **4** Residual current devices (RCDs):

Where RCDs are fitted within an installation a suitable permanent notice durably marked in legible type and no smaller than the example in BS 7671 should be permanently fixed in a prominent position at or near the main distribution board.

> This installation, or part of it, is protected by a device which automatically switches off the supply if an earth fault develops. Test quarterly by pressing the button marked 'T' or 'Test'. The device should switch off the supply and should then be switched on to restore the supply. If the device does not switch off the supply when the button is pressed, seek expert advice.

5 Caravan installations: 721.514.1

All caravans and motor caravans should have a notice fixed near the main switch giving instructions on the connection and disconnection of the caravan installation to the electricity supply.

The notice should be of a durable material permanently fixed, and bearing in indelible and easily legible characters the text shown in BS 7671.

INSTRUCTIONS FOR ELECTRICITY SUPPLY

TO CONNECT

1. Before connecting the caravan installation to the mains supply, check that:

    (a) the supply available at the caravan pitch supply point is suitable for the caravan electrical installation and appliances, and

    (b) the voltage and frequency and current ratings are suitable, and

    (c) the caravan main switch is in the OFF position.

    Also, prior to use, examine the supply flexible cable to ensure there is no visible damage or deterioration.

2. Open the cover to the appliance inlet provided at the caravan supply point, if any, and insert the connector of the supply flexible cable.

3. Raise the cover of the electricity outlet provided on the pitch supply point and insert the plug of the supply cable.

THE CARAVAN SUPPLY FLEXIBLE CABLE MUST BE FULLY UNCOILED TO AVOID DAMAGE BY OVERHEATING

4. Switch on at the caravan main isolating switch.

5. Check the operation of residual current devices (RCDs) fitted in the caravan by depressing the test button(s) and reset.

IN CASE OF DOUBT, OR IF AFTER CARRYING OUT THE ABOVE PROCEDURE THE SUPPLY DOES NOT BECOME AVAILABLE, OR IF THE SUPPLY FAILS, CONSULT THE CARAVAN PARK OPERATOR OR THE OPERATOR'S AGENT OR A QUALIFIED ELECTRICIAN

TO DISCONNECT

6. Switch off at the caravan main isolating switch, unplug the cable first from the caravan pitch supply point and then from the caravan inlet connector.

PERIODIC INSPECTION

Preferably not less than once every three years and annually if the caravan is used frequently, the caravan electrical installation and supply cable should be inspected and tested and a report on their condition obtained as prescribed in BS 7671 Requirements for Electrical Installations published by the Institution of Engineering and Technology and BSI.

514.14.1 **6** Non-standard colours:

If alterations or additions are made to an installation so that some of the wiring complies with the harmonized colours and there is also wiring in the old colours, a warning notice should be affixed at or near the appropriate distribution board with the following wording:

CAUTION

This installation has wiring colours to two versions of BS 7671. Great care should be taken before undertaking extension, alteration or repair that all conductors are correctly identified.

514.15.1 **7** Dual supply:

When an installation includes a generating set that can be employed as an alternative source of supply in parallel with another source, a label having the wording shown below should be fixed at or near:

a the origin of the installation
b the meter position, if remote from the origin
c the consumer unit or distribution board to which the generating set is connected
d all points of isolation of both sources of supply.

WARNING

Isolate both mains and on-site generation before carrying out work

Isolate the mains supply at ....

Isolate the generator at ....

I External influences

611.2 Any known changes in external influences, building structure, and alterations or additions which may have affected the suitability of the wiring for its present load and method of installation should be noted.

Note should also be made of any alterations or additions of an irregular nature to the installation. If unsuitable material has been used, the Report should indicate this together with reference to any evident faulty workmanship or design.

3.10 Periodic testing

3.10.1 Periodic testing general

Periodic testing is supplementary to the inspection of the installation described in Section 3.9.

The same range and level of testing as for an initial inspection and test is not necessarily required, or indeed possible. Installations that have been previously tested and for which there are comprehensive records of test results may not need the same degree of testing as installations for which there are no records of testing.

Periodic testing may cause danger if the correct procedures are not applied. Persons carrying out periodic testing must be competent in the use of the instruments employed and have adequate knowledge and experience of the type of installation being tested in order to prevent, so far as is reasonably practicable, danger.

Sample testing may be carried out, with the percentage being at the discretion of the tester (see Section 3.8.2 and note 2 of Table 3.3).

Wherever a sample test indicates results significantly different from those previously recorded, investigation is necessary. Unless the reason for the difference can be clearly identified as relating only to the sample tested, the size of the sample should be increased. If, in the increased sample, further failures to comply with the requirements of the Regulations are indicated, a 100 per cent test should be made.

3.10.2 Tests to be made

The tests considered appropriate by the person carrying out the inspection should be carried out in accordance with the recommendations in Table 3.3. Reference methods of testing are provided in Section 2.7 of this Guidance Note, but alternative methods which give no less effective results may be used.

▼ **Table 3.3**

Testing to be carried out where practicable on existing installations (see notes 1 and 2)

Test	Recommendation
Protective conductors continuity	Between the earth terminal of distribution boards to the following exposed-conductive-parts: ▶ socket-outlet earth connections (note 4) ▶ accessible exposed-conductive-parts of current-using equipment and accessories (notes 4 and 5)
Bonding conductors continuity	▶ all protective bonding conductors ▶ all necessary supplementary bonding conductors
Ring circuit continuity	Where there are proper records of previous tests, this test may not be necessary. This test should be carried out where inspection/documentation indicate that there may have been changes made to the ring final circuit
Insulation resistance	If tests are to be made: ▶ between live conductors, with line(s) and neutral connected together, and earth at all final distribution boards ▶ at main and sub-main distribution panels, with final circuit distribution boards isolated from mains (note 6)
Polarity	At the following positions: ▶ origin of the installation ▶ distribution boards ▶ accessible socket-outlets ▶ extremity of radial circuits (note 7)
Earth electrode resistance	Test each earth rod or group of rods separately, with the test links removed, and with the installation isolated from the supply source.
Earth fault loop impedance	At the following positions: ▶ origin of the installation ▶ distribution boards ▶ accessible socket-outlets ▶ extremity of radial circuits (note 8)
Functional tests RCDs	Tests as required by Regulation 612.13.1, followed by operation of the functional test button.
Circuit-breakers, isolators and switching devices	Manual operation to prove that the devices disconnect the supply.

Notes to Table 3.3:

1   The person carrying out the testing is required to decide which of the above tests are appropriate by using their experience and knowledge of the installation being inspected and tested and by consulting any available records.

2   Where sampling is applied, the percentage used is at the discretion of the tester. However, a percentage of less than 10 per cent is inadvisable.

3   The tests need not be carried out in the order shown in the table.

4   The earth fault loop impedance test may be used to confirm the continuity of protective conductors at socket-outlets and at accessible exposed-conductive-parts of current-using equipment and accessories.

5   Generally, accessibility may be considered to be within 3 m from the floor or from where a person can stand.

6   Where the circuit includes surge protective devices (SPDs) or other electronic devices which require a connection to earth for functional purposes, these devices will require disconnecting to avoid influencing the test result and to avoid damaging them.

7   Where there are proper records of previous tests, this test may not be necessary.

8   Some earth fault loop impedance testers may trip RCDs in the circuit.

3.10.3   Detailed periodic testing

a   Continuity of protective conductors and equipotential bonding conductors

612.2.1   If an electrical installation is isolated from the supply, it is permissible to disconnect protective and equipotential bonding conductors from the main earthing terminal in order to verify their continuity.

Where an electrical installation cannot be isolated from the supply, the protective and equipotential bonding conductors must NOT be disconnected as, under fault conditions, the exposed and extraneous-conductive-parts could be raised to a dangerous level above earth potential.

The use of an earth fault loop impedance tester is often the most convenient way of continuity testing.

When testing the effectiveness of main equipotential bonding conductors, the resistance value between a service pipe or other extraneous-conductive-part and the main earthing terminal should be of the order of 0.05 Ω or less.

Supplementary bonding conductors should similarly have a resistance of 0.05 Ω or less.

411.3.1.1   It should be confirmed that any exposed-conductive-parts including items of class I equipment such as metal switch plates and luminaires are connected to the earthing arrangements of the installation by a circuit protective conductor.

b   Insulation resistance

Insulation resistance tests should be made on electrically isolated circuits with any electronic equipment which might be damaged by application of the test voltage disconnected or only a measurement to protective earth made, with the line and neutral connected together.

Check that information/warnings are given at the distribution board of circuits or equipment likely to be damaged by testing. Any diagram, chart or table should also include this warning.

Where practicable, the tests may be applied to the whole of the installation with all fuse links in place and all switches closed. Alternatively, the installation may be tested in parts, one distribution board at a time.

A d.c. voltage not less than that stated in Table 2.2 should be employed.

Table 2.2 of this Guide and Table 61 of BS 7671 require a minimum insulation resistance of 1.0 MΩ; however, if the resistance is less than 2 MΩ, further investigation is required to determine the cause of the low reading.

Where equipment is disconnected for these tests and the equipment has exposed-conductive-parts required by the Regulations to be connected to protective conductors, the insulation resistance between the exposed-conductive-parts and all live parts of the equipment should be measured separately and should comply with the requirements of the appropriate British Standard for the equipment. If there is no appropriate British Standard for the equipment, the insulation resistance should be not less than 1.0 MΩ.

c Polarity
Tests should be made to verify that:

1 the polarity is correct at the meter and distribution board
2 every fuse and single-pole control and protective devices are connected in line   612.6
   conductors only
3 conductors are correctly connected to socket-outlets and other accessories/equipment
4 except for E14 and E27 lampholders to BS EN 60238, centre-contact bayonet and Edison screw lampholders have their outer or screw contacts connected to the neutral conductor
5 all multi-pole devices are correctly installed.

It should be established whether there have been any alterations or additions to the installation since its last inspection and test. If there have been no alterations or additions then sample tests should be made of at least 10 per cent of all single-pole and multi-pole control devices and of any centre-contact lampholders, together with 100 per cent of socket-outlets. If any incorrect polarity is found then a full test should be made in that part of the installation supplied by the particular distribution board concerned, and the sample testing increased to 25 per cent for the remainder of the installation. If additional cases of incorrect polarity are found in the 25 per cent sample, a full test of the complete installation should be made.

d Earth fault loop impedance
Where protective measures are used which require a knowledge of earth fault loop impedance, the relevant impedance should be measured, or determined by an equally effective method.

Earth fault loop impedance tests should be carried out at the locations indicated below:

1 origin of the installation
2 distribution boards
3 accessible socket-outlets
4 extremity of radial circuits.

Motor circuits

Loop impedance tests on motor circuits can only be carried out on the supply side of isolated motor controlgear. Continuity tests between the circuit protective conductor and motor are then necessary.

Circuits incorporating an RCD

Where the installation incorporates an RCD, the value of earth fault loop impedance obtained in the test should be related to the rated residual operating current ($I_{\Delta n}$) of the protective device, to verify compliance with Section 413 of BS 7671.

$Z_s\ I_{\Delta n} \leq 50$ V for TN systems

411.5.3 $R_A\ I_{\Delta n} \leq 50$ V for TT systems

e   Operation of residual current devices

612.8.1

411.3.3   Where there is RCD protection, the effective operation of each RCD must be verified by a test simulating an appropriate fault condition independent of any test facility incorporated in the device, followed by operation of the integral test device. The nominal rated tripping current should not exceed 30 mA where the RCD provides additional protection for socket-outlets not exceeding 20 A intended for general use by ordinary persons or mobile equipment having a current value not exceeding 32 A that is being used outdoors.

Refer to Section 2.7.18 for detailed test procedure for RCDs.

f   Operation of overcurrent circuit-breakers

612.13.2   Where protection against overcurrent is provided by circuit-breakers, the manual operating mechanism of each circuit-breaker should be operated to verify that the device opens and closes satisfactorily.

611.3 (viii)   It is not normally necessary or practicable to test the operation of the automatic tripping mechanism of circuit-breakers. Any such test would need to be made at a current substantially exceeding the minimum tripping current in order to achieve operation within a reasonable time. For circuit-breakers to BS EN 60898 a test current of not less than two and a half times the nominal rated tripping current of the device is needed for operation within 1 minute, and much larger test currents are necessary to verify operation of the mechanism for instantaneous tripping.

For circuit-breakers of the sealed type, designed not to be maintained, if there is doubt about the integrity of the automatic mechanism it will normally be more convenient to replace the device than to make further tests. Such doubt may arise from visual inspection, if the device appears to have suffered damage or undue deterioration, or where there is evidence that the device may have failed to operate satisfactorily in service.

Circuit-breakers with the facility for injection testing should be so tested and, if appropriate, relay settings confirmed.

g   Operation of devices for isolation and switching

514.1.1   Where means are provided in accordance with the requirements of the Regulations for isolation and switching, they should be operated to verify their effectiveness and checked to ensure adequate and correct labelling.

Easy access to such devices must be maintained, and effective operation must not be impaired by any material placed near the device. Access and operation areas may be required to be marked to ensure they are kept clear. — 513.1

For isolating devices in which the position of the contacts or other means of isolation is externally visible, visual inspection of operation is sufficient and no testing is required.

The operation of every safety switching device should be checked by operating the device in the manner normally intended to confirm that it performs its function correctly in accordance with the requirements of BS 7671.

Where it is a requirement that the device interrupts all the supply conductors, the use of a test lamp or instrument connected between each line and a connection to neutral on the load side of the switching device may be necessary. Reliance should not be placed on a simple observation that the equipment controlled has ceased to operate.

Where switching devices are provided with detachable or lockable handles in accordance with the Regulations, a check should be made to verify that the handles or keys are not interchangeable with others available within the premises.

Where any form of interlocking is provided, e.g. between a main circuit-breaker and an outgoing switch or isolation device, the integrity of the interlocking must be verified. — 612.13.2

Where switching devices are provided for isolation or for mechanical maintenance switching, the integrity of the means provided to prevent any equipment from being unintentionally or inadvertently energised or reactivated must be verified. — 537.2.1.2 / 537.2.2.3 / 537.3.1.2

Note: An inspection checklist which may be used when carrying out periodic inspections is included in this Guidance Note.

3.10.4 Periodic Inspection Report

BS 7671 requires that the results and extent of periodic inspection and testing shall be recorded on a Periodic Inspection Report and provided to the person ordering the inspection. — 631.2 / 634.1

The report must include: — 631.1

1 a description of the extent of the work, including the parts of the installation inspected and details of what the inspection and testing covered
2 any limitations which may have been imposed during the inspection and testing of the installation
3 details of any damage, deterioration, defects and dangerous conditions and any non-compliance with BS 7671 which may give rise to danger — 634.2
4 schedule of inspections
5 schedule of test results.

Any immediately dangerous condition should preferably be rectified. If not, the defect should be reported in writing without delay to the employer or responsible employee (see Regulation 3 of the Electricity at Work Regulations 1989).

621.2 Installations including those constructed in accordance with earlier editions of BS 7671 should be inspected and tested for compliance with the current edition of BS 7671 and departures recorded. However, reference should be made to the note by the Health and Safety Executive following the preface to BS 7671, that installations conforming to earlier editions and not complying with the current edition do not necessarily fail to achieve conformity with the Electricity at Work Regulations 1989.

Guidance on the action to be taken is to be given in the Observations and Recommendations section of the Periodic Inspection Report by numbering each observation (non-compliance) $\boxed{1}$ to $\boxed{4}$.

If the number $\boxed{1}$ is allocated to an observation, indicating that it requires urgent attention, then the overall assessment must be that it is unsatisfactory. An example of this is an installation which has no earth. If numbers $\boxed{2}$ or $\boxed{3}$ are allocated, the person carrying out the test will have to use judgement to determine whether or not the installation can be classed as satisfactory.

Where an '**X**' (unsatisfactory) observation has been recorded against any item in the schedule of inspections, a corresponding comment should be included in the 'Observations and Recommendations' section.

3.11　Thermographic surveying

Important note: It is recommended that persons refer to the requirements of the Electricity at Work Regulations 1989 and the guidance given in the *Memorandum of Guidance on the Electricity at Work Regulations 1989* (HSR25) prior to undertaking any work activity which places themselves or those under their control in close proximity to live parts.

It is relatively easy to make arrangements to disconnect small installations such as domestic premises from the supply to facilitate periodic inspection and testing.

However, as the size and complexity of an installation increases, isolation of the supply becomes increasingly difficult. This is particularly true where continuity of supply has health implications, as may be the case in hospitals and similar premises, or financial implications as would be the case in banks, share-dealing and commodities markets and the like. Nevertheless, it remains necessary to confirm the continuing suitability of such installations for use. Therefore they must still be subjected to planned and preventative maintenance or regular periodic assessment of their condition.

It may well be possible to carry out a thorough visual inspection of such installations without subjecting the inspector or others in the building to any danger, and such an inspection may identify many common defects caused by use/abuse. Furthermore, experience of such installations may provide a valuable insight into commonly occurring cases of wear and tear.

Some defects cannot be discovered by a visual inspection. For example, incorrectly tightened connections can cause a high resistance joint which can then cause a high temperature to occur locally to the connection. If left uncorrected over time further deterioration of the connection may well occur, resulting in a continuing increase in temperature which may subsequently present a risk of fire. This fire risk will be significantly increased in installations where a build-up of dust or other flammable materials can occur in close proximity to this source of heat. It should also be remembered that increased heat at terminations may also result in accelerated deterioration of the insulation locally. Heating effects symptomatic of a fault or other problem within an electrical installation can also occur as a result of cyclical-load operations, use of conductors of inadequate current-carrying capacity, incorrect load balancing and even more mechanically related issues such as incorrect alignment of motor drive couplings and over-tightened belt-drives.

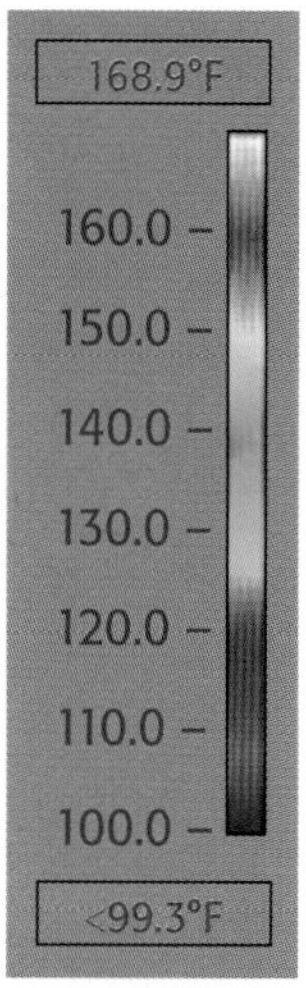

A number of manufacturers offer infra-red imaging equipment which can be used to identify such 'hot spots'. Infra-red thermography works on the principle that all materials emit electromagnetic radiation in the infra-red region which can be detected by a thermal imaging camera.

The amount of radiated energy detected can be presented in a readily usable form, typically being shown as differences in colour that vary with the temperature being detected. Figure 3.1 is an example of a colour/temperature correlation indicator such as those that may accompany images. Such scales will aid the person ordering the inspection or responsible for maintenance activities in their interpretation of the thermal images.

Figure 3.2a shows bolted connections in a busbar as seen by the eye.

In terms of visual inspection, these look to be adequate.

However, if the same connections are viewed using a thermal imaging camera (Figure 3.2b), it is evident that the connections to the centre are running significantly hotter than those to either side.

This higher temperature may indicate a loose connection or connections, but in this case is probably due to the centre bar carrying a significantly higher current than those to each side of it. As such, the person carrying out the inspection could suggest their client looks into improving the load balancing of this part of the installation.

Conversely, with reference to the image of the contactor (Figure 3.3a *overleaf*), it can be seen that the termination to the far right is considerably hotter than the other two.

This is indicative of a loose connection at this termination.

Contrast this with a second image (Figure 3.3b *overleaf*) of the same contactor after the terminal has been suitably tightened. All three terminations are now operating at a much more uniform temperature.

Whilst the remedial work was being carried out it would be sensible to inspect the insulation of the conductors in the terminations to confirm that the insulation remained effective and had not suffered significant damage.

The requirements of the Electricity at Work Regulations 1989 must be taken into account when considering the use of thermographic surveying equipment as its use may necessitate the temporary removal or bypassing of measures that provide basic protection (as defined in BS 7671), such as opening doors to electrical panels and the removal of barriers and covers.

▼ **Figure 3.1**
A colour/temperature correlation indicator

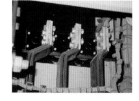

▼ **Figure 3.2a**
Bolted connections in a busbar

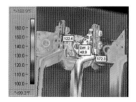

▼ **Figure 3.2b**
Bolted connections in a busbar viewed using thermal imaging

▼ **Figure 3.3a**
Thermal image of the contactor showing termination at far right is too hot

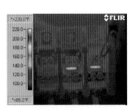

▼ **Figure 3.3b**
Thermal image of the contactor after it has been tightened

The requirements of Regulation 14 (Work on or near live conductors), which is reproduced below, are particularly pertinent:

> **No person shall be engaged in any work activity on or so near any live conductor (other than one suitably covered with insulating material so as to prevent danger) that danger may arise unless –**
>
> **(a) it is unreasonable in all the circumstances for it to be dead; and**
> **(b) it is reasonable in all the circumstances for him to be at work on or near it while it is live; and**
> **(c) suitable precautions (including where necessary the provision of suitable protective equipment) are taken to prevent injury.**

The *Memorandum of Guidance on the Electricity at Work Regulations 1989* (HSR25) recognises that it may be necessary, in some circumstances, for conductors to remain live during testing or diagnostic work. However such work in close proximity to live conductors may only be carried out if it can be done safely and if all precautions required to allow it to be done safely are put in place. Additionally, the work may only be performed by persons who are suitably competent with regard to the type and nature of the work activity being performed as required by Regulation 16 (Persons to be competent to prevent danger or injury). HSR25 also makes clear that although live testing may be justifiable it does not follow that there will necessarily be justification for subsequent repair work to be carried out live.

Persons carrying out thermographic surveying should:

▶ have sufficient competence to prevent danger and injury
▶ understand the system being worked on, the hazards that may arise as a result of the work and the precautions that are required to prevent danger
▶ be able to identify those parts of equipment being inspected which are, or are capable of being live when the supply to the equipment is switched on
▶ implement all precautions required to prevent injury that have been identified as part of the risk assessment for the work
▶ maintain the maximum possible distance from the live or potentially live parts described above at all times
▶ maintain effective control of the area in which the equipment being inspected is situated
▶ ensure that all protective measures which may have been affected by their actions when carrying the inspection work are fully reinstated. All guards and barriers must be replaced and panel doors, lids and covers must be closed and secured properly after the inspection is completed.

Thermographic inspection can be an effective method of identifying defects that would not be located by a more conventional visual inspection. **However, such thermal surveying should not be seen as a substitute for periodic inspection and testing, but rather as an additional tool that can be used by the inspector.** Thermographic surveys can be a highly effective means of targeting preventative maintenance to where it is most required. Defects identified may be factored into the planned maintenance programme for the installation, or where necessary may justify the remedial work being performed without delay.

Test instruments

4.1 Instrument standard

BS EN 61010 *Safety Requirements for Electrical Equipment for Measurement Control and Laboratory Use* is the basic safety standard for electrical test instruments.

The basic instrument standard is BS EN 61557 *Electrical Safety in Low Voltage Distribution Systems up to 1000 V a.c. and 1500 V d.c.* Equipment for testing, measuring or monitoring of protective measures. This standard includes performance requirements and requires compliance with BS EN 61010.

In Section 1.1, reference was made to the use of test leads which conform to HSE Guidance Note GS38. The safety measures and procedures set out in HSE Guidance Note GS38 should be observed for all instruments, leads, probes and accessories. It should be noted that some test instrument manufacturers advise that their instruments be used in conjunction with fused test leads and probes. Other manufacturers advise the use of non-fused leads and probes when the instrument has in-built electrical protection, but it should be noted that such electrical protection does not extend to the probes and leads.

4.2 Instrument accuracy

A basic measurement accuracy of 5 per cent is usually adequate for these test instruments. In the case of analogue instruments, a basic accuracy of 2 per cent of full scale deflection will provide the required accuracy measurement over a useful proportion of the scale.

It should not be assumed that the accuracy of the reading taken in normal field use will be as good as the basic accuracy. The 'operating accuracy' is always worse than the basic accuracy, and additional errors derive from three sources:

1 *Instrument errors:* basic instrument accuracy applies only in ideal conditions; the actual reading accuracy will also be affected by the operator's ability, battery condition, generator cranking speed, ambient temperature and orientation of the instrument .

2 *Loss of calibration:* instruments should be regularly recalibrated using standards traceable to National Standards, or have their accuracy cross-checked using known references such as comparing readings to those obtained from other instruments, or by the use of a proprietary instrument 'check box' having clearly defined characteristics.

   In all cases the type and frequency of recalibration or checking required should be as specified by the instrument manufacturer taking into account ambient environmental and usage factors as appropriate. For example, if an instrument is left in storage at a constant temperature in a dry environment for long periods and is used infrequently, the user may be able to extend the recalibration interval. However, if an instrument is roughly handled and is regularly stored in vehicles and hence is subjected to fluctuations in temperature and humidity caused by changes in time of day/night and time of year then more frequent confirmation of accuracy would be appropriate.

Instruments should also be subjected to regular checks so that errors caused by deterioration of leads, probes connectors etc. do not result in inaccurate readings being recorded when, for example, schedules of test results are compiled.

It is essential that the accuracy of instruments is confirmed after any incidences of mechanical or electrical mishandling.

3 *Field errors:* the instrument reading accuracy will also be affected by external influences as a result of working in the field environment. These influences may be in many forms, and some sources of such inaccuracies are described in the appropriate sections.

BS EN 61557 requires a maximum operating error of ±30% of reading over the stated measurement range.

To achieve satisfactory in-service performance, it is essential to be fully informed about the test equipment, how it is to be used, and the accuracy to be expected.

The accuracy and repeatability of the earth loop impedance measurements at low values is limited by inherent instrument errors and the occurrence of mains transients (and spikes), interference and also close vicinity to distribution transformers. Traditionally in analogue instruments this corresponds to making measurements very near to the zero point on the scale where inaccuracies and non-repeatability are not at all evident with this type of electromechanical movement.

4.3   Low-resistance ohmmeters

612.2.1

The instrument used for low-resistance tests may be either a specialised low-resistance ohmmeter, or the continuity range of an insulation and continuity tester. The test current may be d.c. or a.c. It is recommended that it be derived from a source with no-load voltage between 4 V and 24 V, and a short-circuit current not less than 200 mA.

The measuring range should cover the span 0.2 Ω to 2 Ω, with a resolution of at least 0.01 Ω for digital instruments.

Instruments to BS EN 61557-4 will meet the above requirements.

Field effects contributing to in-service errors are contact resistance, test lead resistance, a.c. interference and thermocouple effects in mixed metal systems.

Whilst contact resistance cannot be eliminated with two-terminal testers, and can introduce errors, the effects of lead resistance can be eliminated by measuring this prior to a test, and subtracting the resistance from the final value. Interference from an external a.c. source (interference pick-up) cannot be eliminated, although it may be indicated by vibration of the pointer of an analogue instrument. Thermocouple effects can be eliminated by reversing the test probes and averaging the resistance readings taken in each direction.

4.4 Insulation resistance ohmmeters

The instrument used should be capable of developing the test voltage required across the load.

The test voltage required is:

1 250 V d.c. for SELV and PELV circuits   Table 61
2 500 V d.c. for all circuits rated up to and including 500 V, but excluding extra-low voltage circuits mentioned above
3 1000 V d.c. for circuits rated above 500 V up to 1000 V.

Instruments conforming to BS EN 61557-2 will fulfil all the above instrument requirements.

The factors affecting in-service reading accuracy include 50 Hz currents induced into cables under test, and capacitance in the test object. These errors cannot be eliminated by test procedures. Capacitance may be as high as 5 µF, and the instrument should have an automatic discharge facility capable of safely discharging such a capacitance. Following an insulation resistance test, the instrument should be left connected until the capacitance within the installation has fully discharged.

4.5 Earth fault loop impedance testers

These instruments operate by circulating a current from the line conductor into the protective earth. This will raise the potential of the protective earth system.

To minimise electric shock hazard from the potential of the protective conductor, the test duration should be within safe limits. This means that the instrument should cut off the test current after 40 ms or a time determined by the safety limits derived from the information contained within DD IEC/TS 60479-1, if the voltage rise of the protective conductor exceeds 50 V during the test.

Instrument accuracy decreases as scale reading reduces. Aspects affecting in-service reading accuracy include transient variations of mains voltage during the test period, mains interference, test lead resistance and errors in impedance measurement as a result of the test method. To allow for the effect of transient voltages the test should be repeated at least once. The other effects cannot be eliminated by test procedures.

For circuits rated up to 50 A, a line–earth loop tester with a resolution of 0.01 Ω should be adequate. In general, such instruments can be relied upon to be accurate down to values of around 0.2 Ω.

Instruments conforming to BS EN 61557-3 will fulfil the above requirements.

These instruments may also offer additional facilities for deriving prospective short-circuit current. The basic measuring principle is generally the same as for earth fault loop impedance testers. The current is calculated by dividing the earth fault loop impedance value into the mains voltage. Instrument accuracy is determined by the same factors as for loop testers. In this case, instrument accuracy decreases as scale reading increases, because the loop value is divided into the mains voltage. It is important to note these aspects, and the manufacturer's documentation should be referred to.

4.6 Earth electrode resistance testers

This may be a four-terminal instrument (or a three-terminal one where a combined lead to the earth electrode would not have a significant resistance compared with the electrode resistance) so that the resistance of the test leads and temporary spike resistance can be eliminated from the test result.

Aspects affecting in-service reading accuracy include the effects of temporary spike resistance, interference currents and the layout of the test electrodes. The instrument should carry some facility to check that the resistance to earth of the temporary potential and current spikes are within the operating limits of the instrument. It may be helpful to note that instruments complying with BS EN 61557-5 incorporate this facility. Care should be exercised to ensure that temporary spikes are positioned with reasonable accuracy.

4.7 RCD testers

The test instrument should be capable of applying the full range of test current to an in-service accuracy as given in BS EN 61557-6. This in-service reading accuracy will include the effects of voltage variations around the nominal voltage of the tester.

To check RCD operation and to minimise danger during the test, the test current should be applied for no longer than 2 s.

Instruments conforming to BS EN 61557-6 will fulfil the above requirements.

4.8 Phase rotation instruments

BS EN 61557-7 gives the requirements for measuring equipment for testing the phase sequence in three-phase distribution systems whether indication is given by mechanical, visual and/or audible means.

BS EN 61557-7 includes requirements that:

▶ Indication shall be unambiguous between 85 per cent and 110 per cent of the nominal system voltage or within the range of the nominal voltage and between 95 per cent and 105 per cent of the nominal system frequency.
▶ The measuring equipment should be suitable for continuous operation.
▶ The measuring equipment should be so designed that when either one or two measuring leads are connected to earth and the remaining measuring lead(s) remain connected to their corresponding phase conductors, the resulting total current to earth should not exceed 3.5 mA rms.
▶ The measuring equipment should not be damaged nor should the user be exposed to danger in situations where the measuring equipment is connected to 120 per cent of the rated system voltage or to 120 per cent of its rated maximum voltage range.
▶ Portable measuring equipment should be provided with permanently attached leads or with a plug device with live parts not accessible, whether plugged or unplugged.

Forms 5

5.1 Initial inspection and testing

Forms 1 to 4 are designed for use when inspecting and testing a new installation, or an alteration or addition to an existing installation. The forms comprise the following:

1 Short form of Electrical Installation Certificate (to be used when one person is responsible for the design, construction, inspection and testing of an installation)
2 Electrical Installation Certificate (Standard form from Appendix 6 of BS 7671)
3 Schedule of Inspections
4 Schedule of Test Results.

Notes on completion and guidance for recipients are provided with the form.

5.2 Minor works

The complete set of forms for initial inspection and testing may not be appropriate for minor works. When an addition to an electrical installation does not extend to the installation of a new circuit, the minor works form may be used. This form is intended for such work as the addition of a socket-outlet or lighting point to an existing circuit, or for repair or modification.

Form 5 is the Minor Electrical Installation Works Certificate from Appendix 6 of BS 7671.

Notes on completion and guidance for recipients are provided with the form.

5.3 Periodic inspection

Form 6, the Periodic Inspection Report from Appendix 6 of BS 7671, is for use when carrying out routine periodic inspection and testing of an existing installation. It is not for use when alterations or additions are made. A Schedule of Inspections (F3) and Schedule of Test Results (F4) should accompany the Periodic Inspection Report (F6).

Notes on completion and guidance for recipients are provided with the form.

5.4 Model forms for certification and reporting

The introduction to Appendix 6 of BS 7671:2008 'Model forms for certification and reporting' is reproduced on this page.

Introduction

(i)   The Electrical Installation Certificate required by Part 6 should be made out and signed or otherwise authenticated by a competent person or persons in respect of the design, construction, inspection and testing of the work.

(ii)   The Minor Works Certificate required by Part 6 should be made out and signed or otherwise authenticated by a competent person in respect of the design, construction, inspection and testing of the minor work.

(iii)   The Periodic Inspection Report required by Part 6 should be made out and signed or otherwise authenticated by a competent person in respect of the inspection and testing of an installation.

(iv)   Competent persons will, as appropriate to their function under (i) (ii) and (iii) above, have a sound knowledge and experience relevant to the nature of the work undertaken and to the technical standards set down in these Regulations, be fully versed in the inspection and testing procedures contained in these Regulations and employ adequate testing equipment.

(v)   Electrical Installation Certificates will indicate the responsibility for design, construction, inspection and testing, whether in relation to new work or further work on an existing installation.

Where design, construction and inspection and testing are the responsibility of one person a Certificate with a single signature declaration in the form shown below may replace the multiple signatures section of the model form.

FOR DESIGN, CONSTRUCTION, INSPECTION & TESTING.

I being the person responsible for the Design, Construction, Inspection & Testing of the electrical installation (as indicated by my signature below), particulars of which are described above, having exercised reasonable skill and care when carrying out the Design, Construction, Inspection & Testing, hereby CERTIFY that the said work for which I have been responsible is to the best of my knowledge and belief in accordance with BS 7671:2008, amended to .... (date) except for the departures, if any, detailed as follows.

(vi)   A Minor Works Certificate will indicate the responsibility for design, construction, inspection and testing of the work described on the certificate.

(vii)   A Periodic Inspection Report will indicate the responsibility for the inspection and testing of an installation within the extent and limitations specified on the report.

(viii) A schedule of inspections and a Schedule of Test Results as required by Part 6 should be issued with the associated Electrical Installation Certificate or Periodic Inspection Report.

(ix)   When making out and signing a form on behalf of a company or other business entity, individuals should state for whom they are acting.

(x)   Additional forms may be required as clarification, if needed by ordinary persons, or in expansion, for larger or more complex installations.

(xi)   The IEE Guidance Note 3 provides further information on inspection and testing on completion and for periodic inspections.

ELECTRICAL INSTALLATION CERTIFICATE
NOTES FOR FORMS 1 AND 2

1 The Electrical Installation Certificate is to be used only for the initial certification of a new installation or for an addition or alteration to an existing installation where new circuits have been introduced.

It is not to be used for a Periodic Inspection for which a Periodic Inspection Report form should be used. For an addition or alteration which does not extend to the introduction of new circuits, a Minor Electrical Installation Works Certificate may be used.

The original Certificate is to be given to the person ordering the work (Regulation 632.1). A duplicate should be retained by the contractor.

2 This Certificate is only valid if accompanied by the Schedule of Inspections and the Schedule(s) of Test Results.

3 The signatures appended are those of the persons authorized by the companies executing the work of design, construction and inspection and testing respectively. A signatory authorized to certify more than one category of work should sign in each of the appropriate places.

4 The time interval recommended before the first periodic inspection must be inserted (see IEE Guidance Note 3 for guidance).

5 The page numbers for each of the Schedules of Test Results should be indicated, together with the total number of sheets involved.

6 The maximum prospective fault current recorded should be the greater of either the short-circuit current or the earth fault current.

7 The proposed date for the next inspection should take into consideration the frequency and quality of maintenance that the installation can reasonably be expected to receive during its intended life, and the period should be agreed between the designer, installer and other relevant parties.

GUIDANCE FOR RECIPIENTS (to be appended to the Certificate)

This safety Certificate has been issued to confirm that the electrical installation work to which it relates has been designed, constructed, inspected and tested in accordance with British Standard 7671 (*IEE Wiring Regulations*).

You should have received an original Certificate and the contractor should have retained a duplicate Certificate. If you were the person ordering the work, but not the owner of the installation, you should pass this Certificate, or a full copy of it including the schedules, immediately to the owner.

The 'original' Certificate should be retained in a safe place and be shown to any person inspecting or undertaking further work on the electrical installation in the future. If you later vacate the property, this Certificate will demonstrate to the new owner that the electrical installation complied with the requirements of British Standard 7671 at the time the Certificate was issued. The Construction (Design and Management) Regulations require that for a project covered by those Regulations, a copy of this Certificate, together with schedules is included in the project health and safety documentation.

For safety reasons, the electrical installation will need to be inspected at appropriate intervals by a competent person. The maximum time interval recommended before the next inspection is stated on Page 1 under 'Next Inspection'.

This Certificate is intended to be issued only for a new electrical installation or for new work associated with an addition or alteration to an existing installation. It should not have been issued for the inspection of an existing electrical installation. A 'Periodic Inspection Report' should be issued for such a periodic inspection.

The Certificate is only valid if a Schedule of Inspections and Schedule of Test Results are appended.

Form 1   Form No   /1

ELECTRICAL INSTALLATION CERTIFICATE (notes 1 and 2)
(REQUIREMENTS FOR ELECTRICAL INSTALLATIONS - BS 7671 [IEE WIRING REGULATIONS])

DETAILS OF THE CLIENT (note 1)

....
....
....

INSTALLATION ADDRESS

....
....
....Postcode ....

DESCRIPTION AND EXTENT OF THE INSTALLATION Tick boxes as appropriate

Description of installation: ....

Extent of installation covered by this Certificate:...
....
....
....
....
....(use continuation sheet if necessary) see continuation sheet No: .....

New installation	☐
Addition to an existing installation	☐
Alteration to an existing installation	☐

FOR DESIGN, CONSTRUCTION, INSPECTION & TESTING
I being the person responsible for the Design, Construction, Inspection & Testing of the electrical installation (as indicated by my signature below), particulars of which are described above, having exercised reasonable skill and care when carrying out the Design, Construction, Inspection & Testing, hereby CERTIFY that the said work for which I have been responsible is to the best of my knowledge and belief in accordance with BS 7671:2008 amended to .... (date) except for the departures, if any, detailed as follows:

> Details of departures from BS 7671 (Regulations 120.3 and 120.4):

The extent of liability of the signatory is limited to the work described above as the subject of this Certificate.

Name (IN BLOCK LETTERS): ....   Position:.......................................
Signature (note 3): ....   Date:...
For and on behalf of:...
Address:..
....
.... Postcode.......................   Tel No: ....

NEXT INSPECTION

I recommend that this installation is further inspected and tested after an interval of not more than .... years/months (notes 4 and 7)

SUPPLY CHARACTERISTICS AND EARTHING ARRANGEMENTS Tick boxes and enter details, as appropriate

Earthing arrangements	Number and Type of Live Conductors	Nature of Supply Parameters	Supply Protective Device Characteristics
TN-C ☐	a.c. ☐ d.c. ☐	Nominal voltage, $U/U_o^{(1)}$ V	Type: ....
TN-S ☐	1-phase, 2-wire ☐ 2-pole ☐	Nominal frequency, $f^{(1)}$ .... Hz	Rated current .... A
TN-C-S ☐	1-phase, 3-wire ☐ 3-pole ☐	Prospective fault current, $I_{pf}^{(2)}$ .... kA (note 6)	
TT ☐	2-phase, 3-wire ☐ other ☐	External loop impedance, $Z_e^{(2)}$........................ Ω	
IT ☐	3-phase, 3-wire ☐	(Note: (1) by enquiry, (2) by enquiry or by measurement)	
Alternative source ☐ of supply (to be detailed on attached schedules)	3-phase, 4-wire ☐		

Page 1 of   (note 5)

PARTICULARS OF INSTALLATION REFERRED TO IN THE CERTIFICATE Tick boxes and enter details, as appropriate

Means of Earthing	**Maximum Demand**
Distributor's facility ☐	Delete as appropriate Maximum demand (load) .... kVA / Amps

Details of Installation Earth Electrode (*where applicable*)

	Type (e.g. rod(s), tape etc)	Location	Electrode resistance to Earth
Installation earth electrode ☐	....	....	.... Ω

Main Protective Conductors

Earthing conductor: material .... csa ....mm² connection verified ☐

Main protective bonding
conductors material .... csa ....mm² connection verified ☐

To incoming water and/or gas service ☐   To other elements ....

Main Switch or Circuit-breaker

BS, Type .... No. of poles .... Current rating ....A Voltage rating V

Location .... Fuse rating or setting ....A

Rated residual operating current $I_{\Delta n}$ = ... mA, and operating time of .... ms (at $I_{\Delta n}$) (Applicable only where an RCD is suitable

COMMENTS ON EXISTING INSTALLATION: (In the case of an alteration or additions see Section 633)

....
....
....
....
....
....
....
....
....
....
....
....
....
....
....
....

SCHEDULES (note 2)
The attached Schedules are part of this document and this Certificate is valid only when they are attached to it.
.... Schedules of Inspections and .... Schedules of Test Results are attached.
(Enter quantities of schedules attached).

Page 2 of   (note 5)

Form 2   Form No   /2

ELECTRICAL INSTALLATION CERTIFICATE (notes 1 and 2)
(REQUIREMENTS FOR ELECTRICAL INSTALLATIONS - BS 7671 [IEE WIRING REGULATIONS])

DETAILS OF THE CLIENT (note 1)
....
....
....

INSTALLATION ADDRESS
....
....
.... Postcode ....

DESCRIPTION AND EXTENT OF THE INSTALLATION Tick boxes as appropriate
(note 1)
Description of installation: ....

Extent of installation covered by this Certificate:..	New installation ☐
....	Addition to an existing installation ☐
....	
....	Alteration to an existing installation ☐
.... (use continuation sheet if necessary) see continuation sheet No:	

FOR DESIGN
I/We being the person(s) responsible for the design of the electrical installation (as indicated by my/our signatures below), particulars of which are described above, having exercised reasonable skill and care when carrying out the design hereby CERTIFY that the design work for which I/we have been responsible is to the best of my/our knowledge and belief in accordance with BS 7671:2008, amended to .... (date) except for the departures, if any, detailed as follows:

> Details of departures from BS 7671 (Regulations 120.3 and 120.4):

The extent of liability of the signatory or the signatories is limited to the work described above as the subject of this Certificate.

For the DESIGN of the installation: **(Where there is mutual responsibility for the design)

Signature: .... Date: .... Name (BLOCK LETTERS):.. Designer No 1

Signature:.............................. Date: .... Name (BLOCK LETTERS):.. Designer No 2**

FOR CONSTRUCTION
I/We being the person(s) responsible for the construction of the electrical installation (as indicated by my/our signatures below), particulars of which are described above, having exercised reasonable skill and care when carrying out the construction hereby CERTIFY that the construction work for which I/we have been responsible is to the best of my/our knowledge and belief in accordance with BS 7671:2008, amended to .... (date) except for the departures, if any, detailed as follows:

> Details of departures from BS 7671 (Regulations 120.3 and 120.4):

The extent of liability of the signatory is limited to the work described above as the subject of this Certificate.

For CONSTRUCTION of the installation:

Signature.. Date..

Name (BLOCK LETTERS) .... Constructor

FOR INSPECTION & TESTING
I/We being the person(s) responsible for the inspection & testing of the electrical installation (as indicated by my/our signatures below), particulars of which are described above, having exercised reasonable skill and care when carrying out the inspection & testing hereby CERTIFY that the work for which I/we have been responsible is to the best of my/our knowledge and belief in accordance with BS 7671:2008, amended to .... (date) except for the departures, if any, detailed as follows:

> Details of departures from BS 7671 (Regulations 120.3 and 120.4):

The extent of liability of the signatory is limited to the work described above as the subject of this Certificate.

For INSPECTION AND TEST of the installation:

Signature.. Date..

Name (BLOCK LETTERS) .... Inspector

NEXT INSPECTION (notes 4 and 7)
I/We the designer(s), recommend that this installation is further inspected and tested after an interval of not more than .... years/months.

Page 1 of   (note 5)

PARTICULARS OF SIGNATORIES TO THE ELECTRICAL INSTALLATION CERTIFICATE (note 3)

Designer (No 1)

Name: .... Company: ....

Address: ....

.... Postcode: .... Tel No: ....

Designer (No 2)
(if applicable)

Name: .... Company: ....

Address: ....

.... Postcode: .... Tel No: ....

Constructor

Name: .... Company: ....

Address: ....

.... Postcode: .... Tel No: ....

Inspector

Name: .... Company: ....

Address: ....

.... Postcode: .... Tel No: ....

SUPPLY CHARACTERISTICS AND EARTHING ARRANGEMENTS Tick boxes and enter details, as appropriate

Earthing arrangements	Number and Type of Live Conductors	Nature of Supply Parameters	Supply Protective Device Characteristics
TN-C ☐	a.c. ☐ d.c. ☐	Nominal voltage, U/U_o [1] V	Type: ....
TN-S ☐			....
TN-C-S ☐	1-phase, 2-wire ☐ 2-pole ☐	Nominal frequency, f [1] .... Hz	Rated current .... A
TT ☐	1-phase, 3-wire ☐ 3-pole ☐	Prospective fault current, I_{pf} [2] .... kA (note 6)	
IT ☐	2-phase, 3-wire ☐ other ☐	External loop impedance, Z_e [2]........................ Ω	
	3-phase, 3-wire ☐	(Note: (1) by enquiry, (2) by enquiry or by measurement)	
Alternative source ☐ of supply (to be detailed on attached schedules)	3-phase, 4-wire ☐		

PARTICULARS OF INSTALLATION REFERRED TO IN THE CERTIFICATE Tick boxes and enter details, as appropriate

Means of Earthing	Maximum Demand Delete as appropriate
Distributor's facility ☐	Maximum demand (load) .... kVA / Amps

Details of Installation Earth Electrode (where applicable)

	Type	Location	Electrode resistance to Earth
Installation earth electrode ☐	(e.g. rod(s), tape etc)		
	....	....	.... Ω

Main Protective Conductors

Earthing conductor: material .... csa ....mm^2 connection verified ☐

Main protective bonding conductors material .... csa ....mm^2 connection verified ☐

To incoming water and/or gas service ☐   To other elements ....

Main Switch or Circuit-breaker

BS, Type .... No. of poles ....   Current rating ....A   Voltage rating V

Location...   Fuse rating or setting ....A

Rated residual operating current $I_{\Delta n}$ = ... mA, and operating time of .... ms (at $I_{\Delta n}$) (applicable only where an RCD is suitable and is used as a main circuit-breaker)

COMMENTS ON EXISTING INSTALLATION: (In the case of an alteration or additions see Section 633)

....

....

....

....

....

....

....

....

SCHEDULES (note 2)
The attached Schedules are part of this document and this Certificate is valid only when they are attached to it.
.... Schedules of Inspections and .... Schedules of Test Results are attached.
(Enter quantities of schedules attached).

Form 3   Form No   /3

SCHEDULE OF INSPECTIONS

Methods of protection against electric shock

Both basic and fault protection:

☐ (i)   SELV (Note 1)

☐ (ii)   PELV

☐ (iii) Double insulation (Note 2)

☐ (iv)   Reinforced insulation (Note 2)

Basic protection: (Note 3)

☐ (i)   Insulation of live parts

☐ (ii)   Barriers or enclosures

☐ (iii) Obstacles (Note 4)

☐ (iv)   Placing out of reach (Note 5)

Fault protection:

(i)   Automatic disconnection of supply:

☐   Presence of earthing conductor

☐   Presence of circuit protective conductors

☐   Presence of protective bonding conductors

☐   Presence of supplementary bonding conductors

☐   Presence of earthing arrangements for combined protective and functional purposes

☐   Presence of adequate arrangements for alternative source(s), where applicable

☐   FELV

☐   Choice and setting of protective and monitoring devices (for fault and/or overcurrent protection)

(ii) Non-conducting location: (Note 6)

☐   Absence of protective conductors

(iii) Earth-free local equipotential bonding: (Note 6)

☐   Presence of earth-free local equipotential bonding

(iv) Electrical Separation: (Note 7)

☐   Provided for **one item** of current-using equipment

☐   Provided for **more than one item** of current-using equipment

Additional protection:

☐   Presence of residual current devices(s)

☐   Presence of supplementary bonding conductors

Prevention of mutual detrimental influence

☐ (a)   Proximity of non-electrical services and other influences

☐ (b)   Segregation of Band I and Band II circuits or use of Band II insulation

☐ (c)   Segregation of safety circuits

Identification

☐ (a)   Presence of diagrams, instructions, circuit charts and similar information

☐ (b)   Presence of danger notices and other warning notices

☐ (c)   Labelling of protective devices, switches and terminals

☐ (d)   Identification of conductors

Cables and conductors

☐   Selection of conductors for current-carrying capacity and voltage drop

☐   Erection methods

☐   Routing of cables in prescribed zones

☐   Cables incorporating earthed armour or sheath, or run within an earthed wiring system, or otherwise adequately protected against nails, screws and the like

☐   Additional protection provided by 30 mA RCD for cables in concealed walls (where required in premises not under the supervision of a skilled or instructed person)

☐   Connection of conductors

☐   Presence of fire barriers, suitable seals and protection against thermal effects

General

☐   Presence and correct location of appropriate devices for isolation and switching

☐   Adequacy of access to switchgear and other equipment

☐   Particular protective measures for special installations and locations

☐   Connection of single-pole devices for protection or switching in line conductors only

☐   Correct connection of accessories and equipment

☐   Presence of undervoltage protective devices

☐   Selection of equipment and protective measures appropriate to external influences

☐   Selection of appropriate functional switching devices

Inspected by ....   Date ....

Notes:

✓   to indicate an inspection has been carried out and the result is satisfactory

✗   to indicate an inspection has been carried out and the result is not satisfactory (applicable to a periodic inspection only)

N/A to indicate the inspection is not applicable to a particular item

LIM to indicate that, exceptionally, a limitation agreed with the person ordering the work prevented the inspection or test being carried out (applicable to a periodic inspection only).

1. SELV – an extra-low voltage system which is electrically separated from Earth and from other systems in such a way that a single-fault cannot give rise to the risk of electric shock. The particular requirements of the Regulations must be checked (see Section 414)

2. Double or reinforced insulation. Not suitable for domestic or similar installations if it is the sole protective measure (see 412.1.3)

3. Basic protection – will include measurement of distances where appropriate

4. Obstacles – only adopted in special circumstances (see 417.2)

5. Placing out of reach – only adopted in special circumstances (see 417.3)

6. Non-conducting locations and Earth-free local equipotential bonding – these are not recognised for general application. May only be used where the installation is controlled/under the supervision of skilled or instructed persons (see Section 418)

7. Electrical separation – the particular requirements of the Regulations must be checked. If a single item of current-using equipment is supplied from a single source, see Section 413. If more than one item of current-using equipment is supplied from a single source then the installation must be controlled/under the supervision of skilled or instructed persons, see also Regulation 418.3

Page   of

Form 4
SCHEDULE OF TEST RESULTS

Contractor: ....

Test Date: ....

Signature ....

Method of fault protection: ....

Equipment vulnerable to testing: ....

Address/Location of distribution board:

....

....

*1 Type of Supply: TN-S/TN-C-S/TT

*2 Ze at origin: ....ohms

*3 PFC: ....kA

Confirmation of supply polarity ☐

Instruments

loop impedance: ....

continuity: ....

insulation: ....

RCD tester: ....

Description of Work: ....

Circuit Description	Overcurrent Device *4 Short-circuit capacity: ......kA		Wiring Conductors		Continuity		Insulation Resistance		Polarity	Earth Loop Imped-ance	Functional Testing		Remarks
	type	Rating I_n	live	cpc	$(R_1 + R_2)^*$	R_2^*	Live/ Live	Live/ Earth		Z_s	RCD time	Other	
		A	mm^2	mm^2	Ω	Ω	$M\Omega$	$M\Omega$	*11	Ω	ms		
1	2	3	4	5	*6	*7	*9	*10		*12	*13	*14	15

Test Results

Deviations from Wiring Regulations and special notes:

*Number - See notes on schedule of test results   **Complete column 6 or 7**

Page   of

SCHEDULE OF TEST RESULTS
NOTES

*1 **Type of supply** is ascertained from the supply company or by inspection.

*2 **Z_e at origin.** When the maximum value declared by the distributor is used, the effectiveness of the earth must be confirmed by a test. If measured the main bonding will need to be disconnected for the duration of the test.

*3 **Prospective fault current (PFC).** The value recorded is the greater of either the short-circuit current or the earth fault current. Preferably determined by enquiry of the distributor.

*4 **Short-circuit capacity** of the device is noted, see Table 7.2.6(i) of the *On-Site Guide* or 2.7.16 of GN3.

The following tests, where relevant, shall be carried out in the following sequence:

Continuity of protective conductors, including main and supplementary bonding

Every protective conductor, including main and supplementary bonding conductors, should be tested to verify that it is continuous and correctly connected.

*6 **Continuity**

Where Test Method 1 is used, enter the measured resistance of the line conductor plus the circuit protective conductor ($R_1 + R_2$).

See 10.3.1 of the *On-Site Guide* or 2.7.5 of GN3.

During the continuity testing (Test Method 1) the following polarity checks are to be carried out:

(a) every fuse and single-pole control and protective device is connected in the line conductor only

(b) centre-contact bayonet and Edison screw lampholders have outer contact connected to the neutral conductor

(c) wiring is correctly connected to socket-outlets and similar accessories.

Compliance is to be indicated by a tick in polarity column 11.

($R_1 + R_2$) need not be recorded if R_2 is recorded in column 7.

*7 Where Test Method 2 is used, the maximum value of R_2 is recorded in column 7.

See 10.3.1 of the *On-Site Guide* or 2.7.5 of GN3.

*8 **Continuity of ring final circuit conductors**

A test shall be made to verify the continuity of each conductor including the protective conductor of every ring final circuit.

See 10.3.2 of the *On-Site Guide* or 2.7.6 of GN3.

*9, *10 **Insulation Resistance**

All voltage sensitive devices to be disconnected or test between live conductors (line and neutral) connected together and earth.

The insulation resistance between live conductors is to be inserted in column 9.

The minimum insulation resistance values are given in Table 10.1 of the *On-Site Guide* or Table 2.2 of GN3.

See 10.3.3 of the *On-Site Guide* or 2.7.7 of GN3.

All the preceding tests should be carried out before the installation is energised.

*11 Polarity

A satisfactory polarity test may be indicated by a tick in column 11.

Only in a Schedule of Test Results associated with a Periodic Inspection Report is it acceptable to record incorrect polarity.

*12 Earth fault loop impedance Z_s

This may be determined either by direct measurement at the furthest point of a live circuit or by adding $(R_1 + R_2)$ of column 6 to Z_e. Z_e is determined by measurement at the origin of the installation or preferably the value declared by the supply company used.

$Z_s = Z_e + (R_1 + R_2)$. Z_s should be less than the values given in Appendix 2 of the *On-Site Guide* or Appendix B of GN3.

*13 Functional testing

The operation of RCDs (including RCBOs) shall be tested by simulating a fault condition, independent of any test facility in the device.

Record operating time in column 13. Effectiveness of the test button must be confirmed. See Section 11 of the *On-Site Guide* or 2.7.15 and 2.7.18 of GN3.

*14 All switchgear and controlgear assemblies, drives, control and interlocks, etc. must be operated to ensure that they are properly mounted, adjusted, and installed.

Satisfactory operation is indicated by a tick in column 14.

Earth electrode resistance

The earth electrode resistance of TT installations must be measured, and normally an RCD is required.

For reliability in service the resistance of any Earth electrode should be below 200 Ω. Record the value on Form 1, 2 or 6, as appropriate. See 10.3.5 of the *On-Site Guide* or 2.7.12 of GN3.

MINOR ELECTRICAL INSTALLATION WORKS CERTIFICATE
NOTES ON COMPLETION

Scope

The Minor Works Certificate is intended to be used for additions and alterations to an installation that do not extend to the provision of a new circuit. Examples include the addition of a socket-outlet or a lighting point to an existing circuit, the relocation of a light switch etc. This Certificate may also be used for the replacement of equipment such as accessories or luminaires, but not for the replacement of distribution boards or similar items. Appropriate inspection and testing, however, should always be carried out irrespective of the extent of the work undertaken.

Part 1 Description of minor works

1,2 The minor works must be so described that the work that is the subject of the certification can be readily identified.

4 See Regulations 120.3 and 120.4. No departures are to be expected except in most unusual circumstances. See also Regulation 633.1.

Part 2 Installation details

2 The method of fault protection must be clearly identified, e.g. Automatic Disconnection of Supply (ADS).

4 If the existing installation lacks either an effective means of earthing or adequate main equipotential bonding conductors, this must be clearly stated. See Regulation 633.2.

Recorded departures from BS 7671 may constitute non-compliance with the Electricity Safety, Quality and Continuity Regulations 2002 (as amended) or the Electricity at Work Regulations 1989. It is important that the client is advised immediately in writing.

Part 3 Essential tests

The relevant provisions of Part 6 'Inspection and testing' of BS 7671 must be applied in full to all minor works. For example, where a socket-outlet is added to an existing circuit it is necessary to:

1 establish that the earthing contact of the socket-outlet is connected to the main earthing terminal

2 measure the insulation resistance of the circuit that has been added to, and establish that it complies with Table 61 of BS 7671

3 measure the Earth fault loop impedance to establish that the maximum permitted disconnection time is not exceeded

4 check that the polarity of the socket-outlet is correct

5 (if the work is protected by an RCD) verify the effectiveness of the RCD.

Part 4 Declaration

1,3 The Certificate shall be made out and signed by a competent person in respect of the design, construction, inspection and testing of the work.

1,3 The competent person will have a sound knowledge and experience relevant to the nature of the work undertaken and to the technical standards set down in BS 7671, be fully versed in the inspection and testing procedures contained in the Regulations and employ adequate testing equipment.

2 When making out and signing a form on behalf of a company or other business entity, individuals shall state for whom they are acting.

Form 5   Form No   /5

MINOR ELECTRICAL INSTALLATION WORKS CERTIFICATE
(REQUIREMENTS FOR ELECTRICAL INSTALLATIONS - BS 7671 [IEE WIRING REGULATIONS])
To be used only for minor electrical work which does not include the provision of a new circuit

---

PART 1: Description of minor works

1. Description of the minor works: ....

2. Location/Address: ....

3. Date minor works completed: ....

4. Details of departures, if any, from BS 7671:

   ....

   ....

   ....

PART 2: Installation details

1. System earthing arrangement: TN-C-S ☐   TN-S ☐   TT ☐

2. Method of fault protection: ....

3. Protective device for the modified circuit: Type BS ....   Rating .... A

4. Comments on existing installation, including adequacy of earthing and bonding arrangements (see Regulation 131.8):

   ....

   ....

   ....

PART 3: Essential Tests

1. Earth continuity satisfactory   ☐

2. Insulation resistance:

   Line/neutral .... $M\Omega$

   Line/earth .... $M\Omega$

   Neutral/earth .... $M\Omega$

3. Earth fault loop impedance Ω

4. Polarity satisfactory   ☐

5. RCD operation (if applicable): rated residual operating current $I_{\Delta n}$ ....mA and operating time of ....ms (at $I_{\Delta n}$)

PART 4: Declaration

1. I/We CERTIFY that the said works do not impair the safety of the existing installation, that the said works have been designed, constructed, inspected and tested in accordance with BS 7671:2008 (IEE Wiring Regulations), amended to ....(date) and that the said works, to the best of my/our knowledge and belief, at the time of my/our inspection, complied with BS 7671 except as detailed in Part 1 above.

2. Name: ....

3. Signature: ....

   For and on behalf of: ....

   Position: ....

   Address: ....

   ....

   Date: ....

   ....

MINOR ELECTRICAL INSTALLATION WORKS CERTIFICATE
GUIDANCE FOR RECIPIENTS (to be appended to the Certificate)

This safety Certificate has been issued to confirm that the electrical installation work to which it relates has been designed, constructed, inspected and tested in accordance with British Standard 7671 (*IEE Wiring Regulations*).

You should have received an original Certificate and the contractor should have retained a duplicate Certificate. If you were the person ordering the work, but not the owner of the installation, you should pass this Certificate, or a full copy of it, immediately to the owner.

A separate Certificate should have been received for each existing circuit on which minor works have been carried out. This Certificate is not appropriate if you requested the contractor to undertake more extensive installation work, for which you should have received an Electrical Installation Certificate.

The 'original' Certificate should be retained in a safe place and be shown to any person inspecting or undertaking further work on the electrical installation in the future. If you later vacate the property, this Certificate will demonstrate to the new owner that the minor electrical installation work carried out complied with the requirements of British Standard 7671 at the time the Certificate was issued.

PERIODIC INSPECTION REPORT
NOTES

1   This Periodic Inspection Report form shall only be used for the reporting on the condition of an existing installation.

2   The Report, normally comprising at least four pages, shall include schedules of both the inspection and the test results. Additional sheets of test results may be necessary for other than a simple installation. The page numbers of each sheet shall be indicated, together with the total number of sheets involved. The Report is only valid if a Schedule of Inspections and a Schedule of Test Results are appended.

3   The intended purpose of the Periodic Inspection Report shall be identified, together with the recipient's details in the appropriate boxes.

4   The maximum prospective fault current recorded should be the greater of either the short-circuit current or the Earth fault current.

5   The 'Extent and Limitations' box shall fully identify the elements of the installation that are covered by the report and those that are not, this aspect having been agreed with the client and other interested parties before the inspection and testing is carried out.

6   The recommendation(s), if any, shall be categorised using the numbered coding 1–4 as appropriate.

7   The 'Summary of the Inspection' box shall clearly identify the condition of the installation in terms of safety.

8   Where the periodic inspection and testing has resulted in a satisfactory overall assessment, the time interval for the next periodic inspection and testing shall be given. The IEE Guidance Note 3 provides guidance on the maximum interval between inspections for various types of buildings. If the inspection and testing reveal that parts of the installation require urgent attention, it would be appropriate to state an earlier re-inspection date, having due regard to the degree of urgency and extent of the necessary remedial work.

9   If the space available on the model form for information on recommendations is insufficient, additional pages shall be provided as necessary.

Form 6   Form No   /6

PERIODIC INSPECTION REPORT FOR AN ELECTRICAL INSTALLATION (note 1)
(REQUIREMENTS FOR ELECTRICAL INSTALLATIONS - BS 7671 [IEE WIRING REGULATIONS])

DETAILS OF THE CLIENT
Client: ....
Address: ....
Purpose for which this Report is required: ....(note 3)

DETAILS OF THE INSTALLATION Tick boxes as appropriate

Occupier: ....

Installation: ....

Address: ....

Description of Premises: Domestic ☐   Commercial ☐   Industrial ☐   Other ☐

....

Estimated age of the Electrical   .... years
Installation:

Evidence of Alterations or Additions: Yes ☐   No ☐   Not apparent ☐

If "Yes", estimate age: .... years

Date of last inspection: ....   Records available   Yes ☐   No ☐

EXTENT AND LIMITATIONS OF THE INSPECTION (note 5)

Extent of electrical installation covered by this report: ....

....

....

....

Limitations: (see Regulation 634.2)...

....

....

This inspection has been carried out in accordance with BS 7671 : 2008 (IEE Wiring Regulations),
amended to .... Cables concealed within trunking and conduits, or cables and conduits concealed under floors,
in roof spaces and generally within the fabric of the building or underground have not been inspected.

NEXT INSPECTION (note 8)

I/We recommend that this installation is further inspected and tested after an interval of not more than ....
months/years, provided that any observations 'requiring urgent attention' are attended to without delay.

DECLARATION

INSPECTED AND TESTED BY

Name: ....   Signature: ....

For and on behalf of: ....   Position: ....

Address: ....

....   Date: ....

....

Page 1 of

SUPPLY CHARACTERISTICS AND EARTHING ARRANGEMENTS Tick boxes and enter details, as appropriate

Earthing arrangements	Number and Type of Live Conductors	Nature of Supply Parameters	Supply Protective Device Characteristics
TN-C ☐ TN-S ☐ TN-C-S ☐ TT ☐ IT ☐ Alternative source ☐ of supply (to be detailed on attached schedules)	a.c. ☐ d.c. ☐ 1-phase, 2-wire ☐ 2-pole ☐ 1-phase, 3 wire ☐ 3-pole ☐ 2-phase, 3-wire ☐ other ☐ 3-phase, 3-wire ☐ 3-phase, 4-wire ☐	Nominal voltage, U/U_o [1] .... V Nominal frequency, f [1] .... Hz Prospective fault current, I_{pf} [2] .... kA (note 4) External loop impedance, Z_e [2] .... Ω *(Note: (1) by enquiry, (2) by enquiry or by measurement)*	Type:................... .... Rated current: ....A

PARTICULARS OF INSTALLATION REFERRED TO IN THE REPORT Tick boxes and enter details, as appropriate

Means of Earthing	Details of Installation Earth Electrode (*where applicable*)		
Distributor's facility ☐ Installation earth electrode ☐	Type (e.g. rod(s), tape etc)	Location	Electrode resistance to Earth .... Ω

Main Protective Conductors

Earthing conductor: material .... csa ....mm² connection verified ☐
Main equipotential bonding conductors material .... csa ....mm² connection verified ☐

To incoming water service ☐   To incoming gas service ☐   To incoming oil service ☐   To structural steel ☐
To lightning protection ☐   To other incoming service(s) ☐   (state details..)

Main Switch or Circuit-breaker

BS, Type... No. of poles ....   Current rating ....A   Voltage rating V

Location ....   Fuse rating or setting ....A

Rated residual operating current $I_{\Delta n}$ = ... mA, and operating time of .... ms (at $I_{\Delta n}$) (applicable only where an RCD is suitable and is used as a main circuit-breaker)

OBSERVATIONS AND RECOMMENDATIONS Tick boxes as appropriate	Recommendations as detailed below
(note 9) Referring to the attached Schedule(s) of Inspections and Test Results, and subject to the limitations specified at the Extent and Limitations of the Inspection section ☐ No remedial work is required ☐ The following observations are made:	note 6

....
....
....
....
....
....
....
....
....
....

One of the following numbers, as appropriate, is to be allocated to each of the observations made above to indicate to the person(s) responsible for the installation the action recommended.

☐1 requires urgent attention   ☐2 requires improvement   ☐3 requires further investigation

☐4 does not comply with BS 7671: 2008 amended to .... This does not imply that the electrical installation inspected is unsafe.

SUMMARY OF THE INSPECTION (note 7)

Date(s) of the inspection: ....

General condition of the installation: ....

....
....
....
....

Overall assessment: Satisfactory/Unsatisfactory (note 8)

SCHEDULE(S)

The attached Schedules are part of this document and this Report is valid only when they are attached to it.
.... Schedules of Inspections and .... Schedules of Test Results are attached.
(Enter quantities of schedules attached).

Page 2 of

PERIODIC INSPECTION REPORT
GUIDANCE FOR RECIPIENTS (to be appended to the Report)

This Periodic Inspection Report form is intended for reporting on the condition of an existing electrical installation.

You should have received an original Report and the contractor should have retained a duplicate. If you were the person ordering this Report, but not the owner of the installation, you should pass this Report, or a copy of it, immediately to the owner.

The original Report is to be retained in a safe place and be shown to any person inspecting or undertaking work on the electrical installation in the future. If you later vacate the property, this Report will provide the new owner with details of the condition of the electrical installation at the time the Report was issued.

The 'Extent and Limitations' box should fully identify the extent of the installation covered by this Report and any limitations on the inspection and tests. The contractor should have agreed these aspects with you and with any other interested parties (Licensing Authority, Insurance Company, Building Society etc.) before the inspection was carried out.

The Report should identify any departures from the safety requirements of the current Regulations and any defects, damage or deterioration that affect the safety of the installation for continued use. **For items classified as 'requires urgent attention', the safety of those using the installation may be at risk,** and it is recommended that a competent person undertakes the necessary remedial work without delay.

For safety reasons, the electrical installation will need to be re-inspected at appropriate intervals by a competent person. The maximum time interval recommended before the next inspection is stated in the Report under 'Next Inspection'.

The Report is only valid if a Schedule of Inspections and a Schedule of Test Results are appended.

Resistance of copper and aluminium conductors A

(This appendix is similar to Appendix 9 of the *On-Site Guide.*)

To check compliance with Regulation 434.5.2 and/or Regulation 543.1.3, i.e. to evaluate the equation $S^2 = I^2 t/k^2$, it is necessary to establish the impedances of the circuit conductors to determine the fault current I and hence the protective device disconnection time t.

434.5.2
543.1.3

Fault current $I = U_o/Z_s$

where:

U$_o$ is the nominal voltage to Earth,

Z$_s$ is the earth fault loop impedance, and

$Z_s = Z_e + R_1 + R_2$

where:

Z$_e$ is that part of the earth fault loop impedance external to the circuit concerned,

R$_1$ is the resistance of the line conductor from the origin of the circuit to the point of utilisation,

R$_2$ is the resistance of the protective conductor from the origin of the circuit to the point of utilisation.

Similarly, in order to design circuits for compliance with BS 7671 and the limiting values of earth fault loop impedance given in Tables 41.2, 41.3 and 41.4 of BS 7671, it is necessary to establish the relevant impedances of the circuit conductors concerned at their operating temperature.

Table A.1 gives values of ($R_1 + R_2$) per metre for various combinations of conductors up to and including 50 mm^2 cross-sectional area. It also gives values of resistance (milliohms) per metre for each size of conductor. These values are at 20 °C.

▼ **Table A.1**
Values of resistance/metre for copper and aluminium conductors and of $R_1 + R_2$ per metre at 20 °C in milliohms/metre

Cross-sectional area (mm²)		Resistance/metre or $(R_1 + R_2)$/metre (mΩ/m)	
Line conductor	Protective conductor	Copper	Aluminium
1	—	18.10	
1	1	36.20	
1.5	—	12.10	
1.5	1	30.20	
1.5	1.5	24.20	
2.5	—	7.41	
2.5	1	25.51	
2.5	1.5	19.51	
2.5	2.5	14.82	
4	—	4.61	
4	1.5	16.71	
4	2.5	12.02	
4	4	9.22	
6	—	3.08	
6	2.5	10.49	
6	4	7.69	
6	6	6.16	
10	—	1.83	
10	4	6.44	
10	6	4.91	
10	10	3.66	
16	—	1.15	1.91
16	6	4.23	
16	10	2.98	—
16	16	2.30	—
			3.82
25	—	0.727	1.20
25	10	2.557	
25	16	1.877	—
25	25	1.454	—
35	—	0.524	2.40
35	16	1.674	0.87
35	25	1.251	2.78
35	35	1.048	2.07
			1.74
50	—	0.387	0.64
50	25	1.114	1.84
50	35	0.911	1.51
50	50	0.774	1.28

Expected ambient temperature (°C)	Correction factor*
0	0.92
5	0.94
10	0.96
15	0.98
20	1.00
30	1.04
40	1.08

▼ **Table A.2**
Ambient temperature multipliers (α) to Table A.1

* The correction factor is given by: {1 + 0.004 (ambient temp − 20 °C)} where 0.004 is the simplified resistance coefficient per °C at 20 °C given by BS 6360 for copper and aluminium conductors.

For verification purposes the designer will need to give the values of the line and circuit protective conductor resistances at the ambient temperature expected during the tests. This may be different from the reference temperature of 20 °C used for Table A.1. The correction factors in Table A.2 may be applied to the Table A.1 values to take account of the ambient temperature (for test purposes only).

A.1 Standard overcurrent devices

Table A.3 gives the multipliers to be applied to the values given in Table A.1 for the purpose of calculating the resistance at maximum operating temperature of the line conductors and/or circuit protective conductors in order to determine compliance with the earth fault loop impedance of Tables 41.2, 41.3 or 41.4 of BS 7671.

Table 41.2
Table 41.3
Table 41.4

Where it is known that the actual operating temperature under normal load conditions is less than the maximum permissible value for the type of cable insulation concerned (as given in the Tables of current-carrying capacity) the multipliers given in Table A.3 may be reduced accordingly.

Multipliers to be applied to Table A.1 for devices in Tables 41.2, 41.3, 41.4

Conductor installation	Conductor insulation		
	70 °C thermoplastic (PVC)	85 °C thermosetting (note 4)	90 °C thermosetting (note 4)
Not incorporated in a cable and not bunched (notes 1, 3)	1.04	1.04	1.04
Incorporated in a cable or bunches (notes 2, 3)	1.20	1.26	1.28

▼ **Table A.3**
Conductor temperature factor F for standard devices

Table 54.2

Table 54.3

Notes:
1 See Table 54.2 of BS 7671. These factors apply when protective conductor is not incorporated or bunched with cables, or for bare protective conductors in contact with cable covering.
2 See Table 54.3 of BS 7671. These factors apply when the protective conductor is a core in a cable or is bunched with cables.
3 The factors are given by F = 1 + 0.004 {conductor operating temperature − 20 °C} where 0.004 is the simplified resistance coefficient per °C at 20 °C given in BS 6360 for copper and aluminium conductors.
4 If cable loading is such that the maximum operating temperature is 70 °C, thermoplastic (70 °C) factors are appropriate.

Table 54.2

Table 54.3

Maximum permissible measured earth fault loop impedance

(The tables in this appendix are as Appendix 2 of the *On-Site-Guide*.)

The tables provide maximum permissible measured earth fault loop impedances (Z_s) for compliance with BS 7671 where conventional final circuits are used. The values are those that must not be exceeded in the tests described at a test ambient temperature of 10 °C to 20 °C. Table B.5 provides correction factors for other ambient temperatures.

612.9

411.4.6
411.4.7
411.4.8
543.1.3

When the cables to be used are to Table 5 and Table 6 of BS6004 or Table 7 of BS7211, or are other thermoplastic (PVC) or lsf cables to these British Standards and if the cable loading is such that the maximum operating temperature is 70 °C, then Tables B.1, B.2 and B.3 give the maximum earth fault loop impedances for circuits:

a with protective conductors of copper and having from 1 mm² to 16 mm² cross-sectional area,

b where the overcurrent protective device is a fuse to BS 88-2, BS 88-6, BS 1361, BS 1362 or BS 3036.

For each type of fuse, two tables are given:

▶ where a disconnection time of 0.4 seconds is required for compliance with Regulation 411.3.2, and

411.3.2

▶ where a disconnection time of 5 seconds is permitted by regulation 411.3.2.3 in a TN system.

411.3.2.3

In each table the earth fault loop impedances given correspond to the appropriate disconnection time from a comparison of the time/current characteristic of the device concerned and the equation given in Regulation 543.1.3.

543.1.3

The tabulated values apply only when the nominal voltage to earth (U_o) is 230 V.

Table B.4 gives the maximum measured Z_s for circuits protected by circuit-breakers to BS EN 60898, RCBOs to BS EN 61009 and circuit-breakers to BS 3871.

Note: The impedances tabulated in this Appendix are maximum measured values at an assumed conductor temperature of 10 °C. These values are lower than those in the tables in BS 7671, which are design figures at the conductor normal operating temperature.

▼ **Table B.1**
Maximum measured earth fault loop impedance (in ohms) when overcurrent protective device is a semi-enclosed fuse to BS 3036 (see note)

Table 41.2
543.1.3

Table 41.4
543.1.3

Table 54.3

(i) 0.4 second disconnection

Protective conductor (mm²)	Fuse rating (amperes)			
	5	15	20	30
1.0	7.7	2.1	1.4	NP
1.5 and above	7.7	2.1	1.4	0.9

(ii) 5 seconds disconnection

Protective conductor (mm²)	Fuse rating (amperes)			
	20	30	45	60
1.0	2.7	NP	NP	NP
1.5	3.1	2.0	NP	NP
2.5	3.1	2.1	1.2	NP
4.0	3.1	2.1	1.3	0.8
6.0 and above	3.1	2.1	1.3	0.9

Note: A value of k of 115 from Table 54.3 of BS 7671 is used. This is suitable for PVC insulated and sheathed cables to Tables 4, 7 or 8 of BS 6004 and for lsf insulated and sheathed cables to Tables 3, 5, 6 or 7 of BS 7211. The k value is based on both the thermoplastic (PVC) and thermosetting (lsf) cables operating at a maximum operating temperature of 70 °C.

NP: protective conductor, fuse combination NOT PERMITTED.

(i) 0.4 second disconnection

Protective conductor (mm²)	Fuse rating (amperes)					
	6	10	16	20	25	32
1.0	6.9	4.1	2.2	1.4	1.2	0.66
1.5	6.9	4.1	2.2	1.4	1.2	0.84
2.5 and above	6.9	4.1	2.2	1.4	1.2	0.84

▼ **Table B.2**
Maximum measured earth fault loop impedance (in ohms) when overcurrent protective device is a fuse to BS 88

Table 41.2
543.1.3

(ii) 5 seconds disconnection

Table 41.4
541.1.3

Protective conductor (mm²)	Fuse rating (amperes)							
	20	25	32	40	50	63	80	100
1.0	1.7	1.2	0.66	NP	NP	NP	NP	NP
1.5	2.3	1.7	1.1	0.64	NP	NP	NP	NP
2.5	2.3	1.8	1.5	0.93	0.55	0.34	NP	NP
4.0	2.3	1.8	1.5	1.1	0.77	0.50	0.23	NP
6.0	2.3	1.8	1.5	1.1	0.84	0.66	0.36	0.22
10.0	2.3	1.8	1.5	1.1	0.84	0.66	0.46	0.33
16.0	2.3	1.8	1.5	1.1	0.84	0.66	0.46	0.34

Note: A value of k of 115 from Table 54.3 of BS 7671 is used. This is suitable for PVC insulated and sheathed cables to Tables 4, 7 or 8 of BS 6004 and for lsf insulated and sheathed cables to Tables 3, 5, 6 or 7 of BS 7211. The k value is based on both the thermoplastic (PVC) and thermosetting (lsf) cables operating at a maximum operating temperature of 70 °C.

NP: protective conductor, fuse combination NOT PERMITTED.

▼ Table B.3

Maximum measured earth fault loop impedance (in ohms) when overcurrent protective device is a fuse to BS 1361

Table 41.2
543.1.3

(i) 0.4 second disconnection

Protective conductor (mm²)	Fuse rating (amperes)			
	5	15	20	30
1.0	8.4	2.6	1.4	0.81
1.5	8.4	2.6	1.4	0.93
2.5 to 16	8.4	2.62	1.4	0.93

Table 41.4
543.1.3

(ii) 5 seconds disconnection

Protective conductor (mm²)	Fuse rating (amperes)					
	20	30	45	60	80	100
1.0	1.7	0.81	NP	NP	NP	NP
1.5	2.2	1.2	0.34	NP	NP	NP
2.5	2.3	1.5	0.52	0.21	NP	NP
4.0	2.3	1.5	0.69	0.37	0.22	NP
6.0	2.3	1.5	0.77	0.53	0.30	0.15
10	2.3	1.5	0.77	0.56	0.40	0.22
16	2.3	1.5	0.77	0.56	0.40	0.29

Table 54.3 **Note:** A value of k of 115 from Table 54.3 of BS 7671 is used. This is suitable for PVC insulated and sheathed cables to Tables 4, 7 or 8 of BS 6004 and for lsf insulated and sheathed cables to Tables 3, 5, 6 or 7 of BS 7211. The k value is based on both the thermoplastic (PVC) and thermosetting (lsf) cables operating at a maximum operating temperature of 70 °C.

NP: protective conductor, fuse combination NOT PERMITTED.

▼ **Table B.4**

▼ **Table B.4** Circuit-breakers. Maximum measured earth fault loop impedance (in ohms) at ambient temperature where the overcurrent device is a circuit breaker to BS 3871 or BS EN 60898 or RCBO to BS EN 61009

For 0.1 to 5 second disconnection times (includes 0.4 and 5 second disconnection times)

Circuit-breaker type	Circuit-breaker rating (amperes)																
	5	6	10	15	16	20	25	30	32	40	45	50	63	100	125
1	9.27	7.73	4.64	3.09	2.90	2.32	1.85	1.55	1.45	1.16	1.03	0.93	0.74	0.46	0.37
2	5.3	4.42	2.65	1.77	1.66	1.32	1.06	0.88	0.83	0.66	0.59	0.53	0.42	0.26	0.21
B	7.42	6.18	3.71	2.47	2.32	1.85	1.48	1.24	1.16	0.93	0.82	0.74	0.59	0.37	0.30
3 & C	6.18	3.09	1.85	1.24	1.16	0.93	0.74	0.62	0.58	0.46	0.41	0.37	0.29	0.19	0.15
D	1.85	1.55	0.93	0.62	0.58	0.46	0.37	0.31	0.29	0.23	0.21	0.19	0.15	0.09	0.07

Note: A value of k of 115 from Table 54.3 of BS 7671 is used. This is suitable for PVC insulated and sheathed cables to Tables 4, 7 or 8 of BS 6004 and for lsf insulated and sheathed cables to Tables 3, 5, 6 or 7 of BS 7211. The k value is based on both the thermoplastic (PVC) and thermosetting (lsf) cables operating at a maximum operating temperature of 70 °C.

NP: protective conductor, fuse combination NOT PERMITTED.

▼ **Table B.5**

Ambient temperature correction factors

Ambient temperature (°C)	Correction factors (from 10 °C) (notes 1, 2)
0	0.96
5	0.98
10	1.00
15	1.02
20	1.04
25	1.06
30	1.08

Notes:

1 The correction factor is given by: {1 + 0.004 (Ambient temp − 10)} where 0.004 is the simplified resistance coefficient per °C at 20 °C given by BS 60228 for both copper and aluminium conductors

2 The factors are different to those of Table A.2 because Table B.5 corrects from 10 °C and Table A.2 from 20 °C. The values in Tables B.1 to B.4 are for a 10 °C ambient temperature.

The appropriate ambient correction factor from Table B.5 is applied to the earth fault loop impedances of Tables B.1 to B.4 if the ambient temperature is not within the range 10 °C to 20 °C.

For example, if the ambient temperature is 25 °C the measured earth fault loop impedance of a circuit protected by a 32 A type 1 circuit-breaker to BS 3871 should not exceed 1.45 x 1.06 = 1.59 Ω.

Appendix 14

BS 7671:2008 introduces a new appendix that provides a means to take into account the increase of the conductor resistance with increase of temperature due to faults which may be used to verify compliance with the requirements of Regulation 411.4 for TN systems.

The requirements of Regulation 411.4.5 are considered met when the measured value of fault loop impedance satisfies the following equation:

$$Z_s(m) \leq 0.8 \times \frac{U_0}{I_a}$$

where:

$Z_s(m)$ is the measured impedance of the fault current loop starting and ending at the point of fault (Ω);

U_0 is the nominal a.c. rms line voltage to earth (V);

I_a is the current causing the automatic operation of the protective device within the time stated in Table 41.1 or within 5 s according to the conditions stated in 411.3.2.3 (A).

B.1 Methods of adjusting tabulated values of Z_s

(See Section 2.7.14 'Verification of test results'.)

A circuit is wired in flat twin and cpc 70 °C thermoplastic (PVC) cable and protected by a 6 amp type B circuit-breaker to BS EN 60898. When tested at an ambient temperature of 20 °C, determine the maximum acceptable measured value of Z_s for the circuit.

Solution:

$Z_{test (max)} = \frac{1}{F} Z_s$

From Table 41.3(a) of BS 7671, the maximum permitted value of Z_s = 7.67 ohms

From Table A.1c in Appendix A of this Guidance Note, F = 1.20

$Z_{test (max)} = \frac{1}{1.20} \times 7.67$

$Z_{test (max)} = 6.39$ ohms

A more accurate value can be obtained if the external earth fault loop impedance, Z_e, is known. In this case, the following formula may be used:

$Z_{test} \leq \frac{1}{F} \{Z_s + Z_e(F - 1)\}$

In the example above, assume Z_e is 0.35. Thus, the more accurate value is:

$Z_{test (max)} = \frac{1}{1.20} \{7.67 + 0.35(1.20 - 1)\}$

$Z_{test (max)} = 6.45$ ohms

Where the test ambient temperature is likely to be other than 20 °C, a further correction can be made to convert the value to the expected ambient temperature, using the following formula:

$Z_{test (max)} = \frac{1}{F + 1 - \alpha} \{Z_s + Z_e(F - \alpha)\}$

where α is given by Table A.2 of Appendix A.

In the example above, assume the ambient temperature is 5 °C.

From Table A.2 in Appendix A of this Guidance Note, α = 0.94

Thus, the accurate reading including temperature compensation is:

$Z_{test (max)} = \frac{1}{1.20 + 1 - 0.94} \{7.67 + 0.35(1.20 - 0.94)\}$

$Z_{test (max)} = 0.794\{7.67 + 0.091\}$

$Z_{test (max)} = 6.16$ ohms

Alternatively, a conductor temperature resistance factor, F, of 1.26, which corresponds to 5 °C, can be used instead of the 1.20 factor in the formula shown in the second solution box. This gives the same result as that shown above.

Note:

If reduced cross-sectional area protective conductors are used, maximum earth fault loop impedances may need to be further reduced to ensure disconnection times are sufficiently short to prevent overheating of protective conductors during earth faults. The requirement of the equation in Regulation 543.1.3 needs to be met:

$$S \geq \frac{\sqrt{I^2 t}}{k}$$

where:

    S   is the nominal cross-sectional area of the conductor in mm^2

    $\geq$ means greater than or equal to

    k   is a factor from Tables 54.2, 54.3 or 54.4

    I   is the prospective earth fault current given by U_{oc}/Z_s

    Z_s is the loop impedance at conductor normal operating temperature

    t   is the operating time of the overcurrent device in seconds. This is obtained from the graphs in Appendix 3 of BS 7671, as the prospective earth fault current I ($= U_{oc}/Z_s$) is known.

The following example illustrates how measurements taken at 20 °C may be adjusted to 70 °C values, taking the $R_1 + R_2$ reading for the circuit into account.

> In the previous example, taking the ($R_1 + R_2$) reading for the circuit as 0.2 ohm:
>
> Z_s for the circuit at 70 °C
>
> $= Z_e + F(R_1 + R_2)_{test}$
>
> $= 0.35 + 1.20 \times 0.2$
>
> $= 0.59$ ohm
>
> The temperature-corrected Z_s figure of 0.59 ohm is acceptable, since it is less than the maximum value of 7.67 ohms given in Table 41.3 of BS 7671.

The formula above involves taking measurements at 20 °C and converting them to 70 °C values. Alternatively, the 70 °C values can be converted to the values at the expected ambient temperature, e.g. 20 °C, when the measurement is carried out.

> Taking the same circuit,
>
> $Z_{test} = Z_e + (R_1 + R_2)_{test}$
>
> $= 0.35 + 0.2$
>
> $= 0.55$ ohm
>
> From the formula $Z_{test\,(max)} = \frac{1}{F} Z_s$
>
> $Z_{s\,(max)}$ from BS 7671 $= 7.67$ ohms
>
> $Z_{test\,(max)} = \frac{1}{1.20} \times 7.67 = 6.39$ ohms
>
> Therefore, as 0.55 is less than 6.39, the circuit is acceptable.

Index

IEE Wiring Regulations and associated publications

The IEE prepares regulations for the safety of electrical installations for buildings, the *IEE Wiring Regulations* (BS 7671 *Requirements for Electrical Installations*), which have now become the standard for the UK and many other countries. It also recommends, internationally, the requirements for ships and offshore installations. The IEE provides guidance on the application of the installation regulations through publications focused on the various activities from design of the installation through to final test and then maintenance. This includes a series of eight Guidance Notes, two Codes of Practice and Model Forms for use in Wiring Installations.

Requirements for Electrical Installations BS 7671:2008 (IEE Wiring Regulations, 17th Edition)
Order book PWR1700B Paperback 2008
ISBN: 978-0-86341-844-0   **£65**

On-Site Guide (BS 7671:2008 17th Edition)
Order book PWGO170B Paperback 2008
ISBN: 978-0-86341-854-9   **£20**

Wiring Matters Magazine   FREE
If you wish to receive a FREE copy or advertise in Wiring Matters please visit
www.theiet.org/wm

IEE Guidance Notes

A series of Guidance Notes has been issued, each of which enlarges upon and amplifies the particular requirements of a part of the IEE Wiring Regulations.

Guidance Note 1: Selection & Erection of Equipment, 5th Edition
Order book PWG1170B 216pp Paperback 2009
ISBN: 978-0-86341-855-6   **£30**

Guidance Note 2: Isolation & Switching, 5th Edition
Order book PWG2170B 74pp Paperback 2009
ISBN: 978-0-86341-856-3   **£25**

Guidance Note 3: Inspection & Testing, 5th Edition
Order book PWG3170B 128pp Paperback 2008
ISBN: 978-0-86341-857-0   **£25**

Guidance Note 4: Protection Against Fire, 5th Edition
Order book PWG4170B 104pp Paperback 2009
ISBN: 978-0-86341-858-7   **£25**

Guidance Note 5: Protection Against Electric Shock, 5th Edition
Order book PWG5170B 115pp Paperback 2009
ISBN: 978-0-86341-859-4   **£25**

Guidance Note 6: Protection Against Overcurrent, 5th Edition
Order book PWG6170B 113pp Paperback 2009
ISBN: 978-0-86341-860-0   **£25**

Guidance Note 7: Special Locations, 3rd Edition
Order book PWG7170B 142pp Paperback 2009
ISBN: 978-0-86341-861-7   **£25**

Guidance Note 8: Earthing and Bonding, 1st Edition
Order book PWRG0241 168pp Paperback 2007
ISBN: 978-0-86341-616-3   **£25**

continues overleaf ▶

Other guidance publications

**Commentary on IEE Wiring Regulations
(17th Edition, BS 7671:2008)**
Order book PWR08640
c.400pp Hardback 2009
ISBN: 978-0-86341-966-9 **£65**

Electrical Maintenance, 2nd Edition
Order book PWR05100
230pp Paperback 2006
ISBN: 978-0-86341-563-0 **£35**

**Code of Practice for In-service Inspection and
Testing of Electrical Equipment, 3rd Edition**
Order book PWR08630
138pp Paperback 2007
ISBN: 978-0-86341-833-4 **£35**

**Electrical Craft Principles, Volume 1,
5th Edition**
Order book PBNS0330
344pp Paperback 2009
ISBN: 978-0-86341-932-4 **£25**

**Electrical Craft Principles, Volume 2,
5th Edition**
Order book PBNS0340
432pp Paperback 2009
ISBN: 978-0-86341-933-1 **£25**

**Electrician's Guide to the Building
Regulations, 2nd Edition**
Order book PWGP170B
234pp Paperback 2008
ISBN: 978-0-86341-862-4 **£20**

**Electrical Installation Design Guide:
Calculations for Electricians and Designers**
Order book PWR05030
186pp Paperback 2008
ISBN: 978-0-86341-550-0 **£20**

Electrician's Guide to Emergency Lighting
Order book PWR05020
88pp Paperback 2009
ISBN: 978-0-86341-551-7 **£20**

Electrical training courses

We offer a comprehensive range of technical training at
many levels, serving your training and career development
requirements as and when they arise.

Courses range from Electrical Basics to Qualifying City &
Guilds or EAL awards.

Train to the 17th Edition BS 7671:2008

- ▶ Update from 16th to 17th Edition
- ▶ Understand the changes
- ▶ New qualifying awards C&G/EAL
- ▶ Meet industry standards

Qualifying Courses

- ▶ Certificate of Competence Management of Electrical
  Equipment Maintenance (PAT) – 1 day
- ▶ Certificate of Competence for the Inspection and
  Testing of Electrical Equipment (PAT) – 1 day
- ▶ Certificate in the Requirements for Electrical
  Installations – 3 days
- ▶ Upgrade from 16th Edition achieved since 2001 –
  1 day
- ▶ Certificate in Fundamental Inspection, Testing and
  Internal Verification – 3 days
- ▶ Certificate in Inspection, Testing and Certification of
  Electrical Installations – 3 days

Other 17th Edition Courses

- ▶ Earthing & Bonding – For designers and electrical
  contractors who require a good working knowledge
  of the E & B arrangements as required by
  BS 7671:2008
- ▶ 17th Edition Design – BS 7671 and the principles
  associated with the design of electrical installations

To view all our current courses and book online, visit
www.theiet.org/coursesbr

**To discuss your training requirements and for on
site group training, please speak to one of our
advisors on +44 (0) 1438 767289**

Collective **inspiration**

Order Form

How to order

BY PHONE:
+44 (0)1438 767328
BY FAX:
+44 (0)1438 767375
BY EMAIL:
sales@theiet.org
BY POST:
The Institution of
Engineering
and Technology,
PO Box 96,
Stevenage
SG1 2SD, UK
OVER THE WEB:
www.theiet.org/books

*Postage/Handling: Postage within the UK is
£3.50 for any number of titles. Outside UK
(Europe) add £5.00 for first title and £2.00 for
each additional book. Rest of World add
£7.50 for the first book and £2.00 for each
additional book. Books will be sent via air-
mail. Courier rates are available on request,
please call +44 (0) 1438 767328 or
email sales@theiet.org for rates.

** To qualify for discounts, member orders
must be placed directly with the IET.

GUARANTEED RIGHT OF RETURN:
If at all unsatisfied, you may return book(s)
in new condition within 30 days for a full
refund. Please include a copy of the invoice.

DATA PROTECTION:
The information that you provide to the IET
will be used to ensure we provide you with
products and services that best meet your
needs. This may include the promotion of
specific IET products and services by post
and/or electronic means. By providing us
with your email address and/or mobile
telephone number you agree that we may
contact you by electronic means. You can
change this preference at any time
by visiting www.theiet.org/my.

Details

Name:

Job Title:

Company/Institution:

Address:

Postcode: Country:

Tel: Fax:

Email:

Membership No (if Institution member):

Payment methods

☐ By **cheque** made payable to The Institution of Engineering and Technology

☐ By **credit/debit card:**

☐ Visa ☐ Mastercard ☐ American Express ☐ Maestro Issue No:_____

Valid from: ☐☐ ☐☐   Expiry Date: ☐☐ ☐☐   Card Security Code: ☐☐☐☐
(3 or 4 digits on reverse of card)

Card No: ☐☐☐☐ ☐☐☐☐ ☐☐☐☐ ☐☐☐☐

Signature_____ Date _____
(Orders not valid unless signed)

Cardholder Name:

Cardholder Address:

Town: Postcode:

Country:

☐ By official **company purchase order** (please attach copy)
EU VAT number:_____

Ordering information

Quantity	Book No.	Title/Author	Price (£)
		Subtotal	
		- Member discount**	
		+ Postage /Handling*	
		+ VAT (if applicable)	
		Total	

The Institution of Engineering and Technology is registered as a Charity in England & Wales (no 211014) and Scotland (no SC038698).

Membership

Passionate about engineering? Committed to your career?

Do you want to join an organisation that is inspiring, insightful and innovative?

One of the most highly recognised knowledge sharing networks in the world, membership to the Institution of Engineering and Technology is for engineers and technologists working or studying in an increasingly multidisciplinary, digital and global environment.

Joining the IET and having access to tailored products and services will become invaluable for your career and can be your first step towards professional qualifications.

You could take advantage of ...

▶ Fortnightly copy of the industry's leading publication, *Engineering & Technology* magazine.

▶ Professional development and career support services to help gain registration.

▶ Dedicated training courses, seminars and events covering a wide range of subjects and skills.

▶ Watch live IET.tv event footage at your desktop via the internet, ask the speaker questions during live streaming and feel part of the audience without physically being there.

▶ Access to over 100 local networks around the world.

▶ Meet like-minded professionals through our array of specialist online communities.

▶ Instant online access to over 70,000 books, 3,000 periodicals and full-text collections of electronic articles – wherever you are in the world.

▶ Discounted rates on IET books and technical proceedings.

Join online today www.theiet.org/join or contact our membership and customer service centre on +44 (0)1438 765678

Professional Registration
What type of registration is for you?

Chartered Engineers (CEng) develop appropriate solutions to engineering problems, using new or existing technologies, through innovation, creativity and change. They might develop and apply new technologies, promote advanced designs and design methods, introduce new and more efficient production techniques, marketing and construction concepts, pioneer new engineering services and management methods. Chartered Engineers are engaged in technical and commercial leadership and possess interpersonal skills.

Incorporated Engineers (IEng) maintain and manage applications of current and developing technology, and may undertake engineering design, development, manufacture, construction and operation. Incorporated Engineers are engaged in technical and commercial management and possess effective interpersonal skills.

Engineering Technicians (EngTech) are involved in applying proven techniques and procedures to the solution of practical engineering problems. You will carry supervisory or technical responsibility, and are competent to exercise creative aptitudes and skills within defined fields of technology. Engineering Technicians also contribute to the design, development, manufacture, commissioning, operation or maintenance of products, equipment, processes or services.

For further information on Professional Registration (CEng/IEng/EngTech), tel: +44 (0)1438 767282 or email: membership@theiet.org

Notes

Notes

Notes

Notes